Introduction to Linear Algebra in Geology

Introduction to Linear Algebra in Geology

J Ferguson

CHAPMAN & HALL

London · Glasgow · Weinheim · New York · Tokyo · Melbourne · Madras

Published by Chapman & Hall, 2-6 Boundary Row, London SE1 8HN, UK

Chapman & Hall, 2-6 Boundary Row, London SE1 8HN, UK

Blackie Academic & Professional, Wester Cleddens Road, Bishopbriggs, Glasgow G64 2NZ, UK

Chapman & Hall Inc., One Penn Plaza, 41st Floor, New York NY 10119, USA

Chapman & Hall Japan, Thomson Publishing Japan, Hirakawacho Nemoto Building, 6F, 1-7-11 Hirakawa-cho, Chiyoda-ku, Tokyo 102, Japan

Chapman & Hall Australia, Thomas Nelson Australia, 102 Dodds Street, South Melbourne, Victoria 3205, Australia

Chapman & Hall India, R. Seshadri, 32 Second Main Road, CIT East, Madras 600 035, India

First edition 1994

© 1994 J. Ferguson

Printed in Great Britain by Page Bros, Norwich

ISBN 0 412 49350 0

Contents

Preface and acknowledgements

Geology is a very visual science and much day-to-day material used as an aid to communication is in the form of maps and sections, or some other type of diagram. Changes in scale or more importantly changes of co-ordinate system are often taken for granted, with little thought being given as to how they are achieved. In petrology, the diagrammatic approach reaches its limits when considering multi-component systems, where attempts to show inter-relationships between more than three components become, to say the least, confusing. Also, in the area of multivariate statistics, much data analysis is carried out because of the ease with which suitable computer software can be obtained and used. Sadly, practice obtained by running such software does not generally lead to an understanding of the methods of multivariate statistics, nor of the limitations of the mathematics behind them.

Linking these three themes – change of coordinate system, interrelating three or more mineral components and multivariate statistics – is the mathematics of linear algebra. Add to this its importance in structural geology, then we have a self-contained mathematical topic worthy of the attention of geologists engaged in most fields of the subject.

Recent experiences teaching 'remedial' mathematics to post-graduate students, has highlighted once again the considerable lack of numeracy amongst some geology graduates of British universities. It showed in particular that it is not safe to make assumptions about basic knowledge of any mathematical topic. This book therefore, develops linear algebra by considering first the mathematics of a straight line. It is hoped that those with some mathematical knowledge are not put off by this simple approach, while others will find some advantage in starting at the beginning. Sadly to save space some necessary mathematical and statistical topics can be mentioned only in passing with suitable references, in the hope that the interested reader will benefit by following them up.

The material in this book is not wholly original and the author acknowledges his debt of gratitude to those mathematicians and geologists who have written on linear algebra in the past. In particular to those authors mentioned in the references, a special thank you for their foresight and clarity. I am also indebted to past students who have patiently listened to lectures, queried points which were not clear and worked through endless examples, since without them this book would never have been written.

Finally I would like to express my gratitude to Dr Glyn Jones and the staff of Technical Communications and especially to my editor Jenny Lawson and her colleagues, whose attention to detail have helped to eliminate a number of potential problems and improve the text. However, in the final analysis I must accept the responsibility for any errors or inaccuracies in the text as published.

Glossary of symbols and conventions

*   Used to denote multiplication in the arithmetic sense.

.   Used to denote multiplication in vector and matrix algebra, when multiplying vectors and matrices together (the dot product Chapter 2).

× Used when indicating the size of a matrix, thus 2 × 2 (read as 2 by 2) indicates a matrix with 2 rows and 2 columns.

The following symbols will be used in matrix algebra:

A scalar is indicated by a Greek letter e.g. psi ψ

A vector is a lower case letter in bold face e.g. **a**

A matrix is an upper case letter in bold face e.g. **B** or an upper case letter in square braces e.g. **[B]**

The transpose of a matrix is $[\mathbf{A}]^T$ or $[\mathbf{A}]'$ (with or without braces)

The determinant of a matrix is det **A** or $|\mathbf{A}|$ (straight lines)

Position of an element in a matrix will be given as i, j (no braces) where i is the row number and j the column number e.g. 2, 2

A vector will always be enclosed in braces e.g. (2 2)

Other symbols will be defined as and when they are needed.

In examples in the text which are taken from the field of silicate geochemistry, the symbols used to denote vectors and matrices relating molecular end-members, chemical analysis, oxides and atoms, will be those proposed by Perry (1967). Since Perry's paper may not be easily accessible to readers, they are summarized here for reference. (Perry, 1967 p.1045)

Matrix **A**   = number of moles of oxide i in molecular member j.

Matrix **D**   = number of atoms k in oxide i.

Matrix **G**   = number of atoms k in molecular member j.

Vector **b**   = mineral chemical analysis vector whose components are the number of moles of oxide b_i in the given mineral.

Vector **w** = structural formulae vector whose components are the number of atoms w_k of a given mineral.

Vector **x**   = composition vector whose components are the number of moles x_j of molecular member j.

Any element of the vector **b** can be calculated as:

$$b_i = \frac{\text{wt. \% of oxide } i}{\text{molecular wt. oxide } i}$$

The vector **w** can be recalculated subject to the restriction that the last element (of the vector), normally oxygen, contains the number of atoms of oxygen corresponding to the silicate group to which the mineral belongs. Thus for example if the mineral were a chlorite the vector could be calculated subject to O_{14}, whereas if the mineral were a hornblende then it would be subject to O_{24}.

Introduction

The power of linear algebra (also referred to as matrix algebra) as an analytical tool in science and engineering has long been accepted. The methods developed have been exploited with increasing frequency in recent years because of the ease of access to desktop computers. Geologists, traditionally slow in applying numerical methods to their subject have nonetheless made much progress in the application of these methods. Applications have covered the fields of structural geology; petrology (in particular metamorphic petrology); petroleum geology (well–log analysis); computer modelling and geostatistics.

In structural geology the use of transformation matrices has found application in the analysis of strain, while in petrology the linear nature of chemical equations has been exploited allowing data to be manipulated according to systems of end-members, whether they be mineral species or oxides of their component elements. Elsewhere, as for example in studies of continental drift, transformations of co-ordinate systems based on traditional algebraic and trigonometrical methods have been rewritten in matrix form, simplifying calculations as well as aiding understanding. Linear algebra has proved to be a particularly powerful tool in multivariate statistics, which has led to its application in analysing multicomponent geochemical data, which is so readily available from modern analytical instruments.

Much of linear algebra is concerned with the generalized properties of the straight line. Thus the simplest linear equation describes the straight line which relates two variables, while the addition of a third variable enables the description of a plane. Although these simple concepts have many applications in geology, we are more frequently concerned with points in space. Such points can be described by the intersection of lines or planes, with reference to a set of mutually orthogonal axes.

In many practical situations such as in geological mapping, we as geologists are concerned with solid bodies whose reference points are defined relative to three mutually orthogonal axes, requiring three spatial variables to describe them. In petrology and geostatistics on the other hand the number of axes (variables) can be large, for example the hornblende series of minerals has ten end-members, involving eleven oxides. Thus in order to describe the hornblende series in terms of the end-members, eleven equations in ten unknowns will be required. Fortunately the rules of the algebra remain the same whether we are dealing with two, three or twenty variables. Easy access to powerful desk-top computers means that the apparently long-winded and tedious computations can be carried out quickly and easily.

The solution of simultaneous linear equations also has an important part to play in many algorithms commonly used in numerical analysis. For example in the area of curve or surface fitting, traditional algebraic techniques lead to sets of simultaneous linear equations which have to be solved. Similarly techniques such as the finite difference method for the solution of differential equations, also involve the solution of simultaneous linear equations. Thus linear algebra should provide algorithms which will allow the accurate solution of such sets of equations. Indeed much of the algebra has developed in response to this problem and has brought together traditional Western and Chinese mathematics.

Linear algebra does have shortcomings, which under certain circumstances can limit its use in practical situations. Therefore the common pitfalls and sources of error will be discussed as appropriate, in order that the student can be aware of them and take appropriate action to avoid their consequences.

Throughout the text simple examples have been chosen to illustrate the basic operations, involving only two or three variables. In order to make the text more relevant for students of geology, geological examples are given to amplify these operations. In these cases the number of variables will vary according to the needs of the problem. Also, in order not to burden the student, many of the formal proofs are omitted, and students who require a more formal approach are recommended to consult one or more of the mathematical texts listed in the references.

Readers should be aware that in a book such as this it is not possible to cover in detail mathematical topics which are not strictly in the field under discussion. It has therefore been assumed that those using this book will be knowledgeable in mathematics at least to the British GCE A level standard, or first year university or college. Finally, since this work is intended primarily for geologists, it is assumed that readers will have a basic working knowledge of the subject, up to at least second year BSc.

For those without the necessary mathematical skills and those who require some revision, there are a number of texts which are useful. Among those which can be recommended are *A Level Mathematics – Course Companion* in the Letts Study Aids Series (Graham *et al.*, 1984). More advanced texts are *Mathematical Methods for Science Students* (Stephenson, 1973), *Mathematical Methods in the Physical Sciences* (Boas, 1983) and *Numerical Methods for Mathematics, Science and Engineering* (Mathews, 1992). Elsewhere in the text other references, both geological and mathematical, are given as required.

As will be noticed, published geological applications of linear algebra have generally followed the standard mathematical conventions with regard to the use of symbols and the labelling of axes. Such established conventions will also be followed in this book. However, where there has been a departure from this practice the geological convention will take precedence unless there is the possibility of misinterpretation. Such instances will be explained as and when they arise.

1

The solution of simultaneous linear equations

In this chapter, methods for the solution of simultaneous linear equations (SLE) based on traditional algebraic methods are examined in some detail. The techniques demonstrated are developments from a straightforward method for the solution of a pair of equations in two unknowns. This means being methodical in the approach to the elimination of one or other of the unknowns, and then extending the method to cope with other more complex situations. As will be shown later these simple row operations can also be used to tackle other useful and necessary applications in linear algebra.

1.1 Basic definitions and introductory examples

In its simplest form a linear equation describes a straight line with reference to a pair of orthogonal axes as illustrated in Figs 1.1a and 1.1b. It defines the slope of the line relative to one of the axes (usually the horizontal), as well as the intercept of the line on the other axis. The general form of the equation expressing the line in the $x_1 x_2$-plane, is:

$$x_2 = a_1 x_1 + z$$

where, a_1 is the slope of the line i.e. the tangent of the measured angle which the line makes with the horizontal, x_1-axis.

If a_1 is positive, the line slopes up from left to right (Fig. 1.1a);

If a_1 is negative the line slopes down from left to right (Fig. 1.1b).

z is the intercept on the vertical, x_2-axis.

If z is positive, the line will cut the x_1-axis with $x_1 < 0$;

If z is zero, then the line passes through the origin;

If z is negative, the line will cut the x_1-axis with $x_1 > 0$.

x_1 and x_2 are any pair of measured variables which are linearly related.

An equation of the general form:

$$x_3 = a_1 x_1 + a_2 x_2 + z$$

will describe a plane surface such as in Figs 1.2a and 1.2b. In this example x_1 and x_2 are the geographical co-ordinates and x_3 some geological

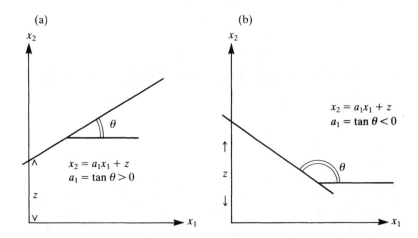

Figure 1.1 Generalized representation of the linear equation for two variables in the x plane. (a) $x_2 = a_1x_1 + z$, $a_1 > 0$; (b) $x_2 = a_1x_1 + z$, $a_1 < 0$.

parameter measured at a number of points defined by x_1 and x_2. Remember that since the line or the surface being considered has only linear terms in its description, the line will be straight and the surface will be flat. A simple three-dimensional geological example would be a map showing the dip and strike of the surface of a flat gently dipping bed (Fig. 1.2b).

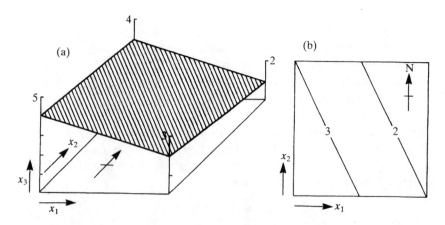

Figure 1.2 Generalized representation of the linear surface $x_3 = a_1x_1 + a_2x_2 + z$: (a) plotted as a surface (shaded); and (b) plotted as a contoured map.

The general form of the equation is;

$$x_n = a_1 x_1 + a_2 x_2 + \cdots + a_m x_m + z$$

Equations of this general form can be written to include as many variables as required.

For any linear equation, provided we are given values for the coefficients $a_1, a_2, \ldots, a_m$ and z, then for any desired values $x_1, x_2 \ldots, x_{n-1}$, we can calculate a corresponding value x_n. Of more importance in many fields of science is the calculation to find points (or lines) of intersection when the equations of two or more lines or surfaces are given. Let us suppose that we have two straight lines defined in terms of a pair of common orthogonal axes:

$$a_{1,1} x_1 + a_{1,2} x_2 = z_1$$
$$a_{2,1} x_1 + a_{2,2} x_2 = z_2$$

where $a_{1,1}, a_{1,2}, a_{2,1}, a_{2,2}, z_1$ and z_2 are given numerical values and the slope of the first line will be $-a_{1,1}/a_{1,2}$ intercept $z_1/a_{1,2}$ and the slope of the second line will be $-a_{2,1}/a_{2,2}$ and intercept $z_2/a_{2,2}$. Notice that the form of the linear equation has been changed from that given earlier, and both x-terms now appear on one side of the equation. A point of intersection will be given by values of x_1 and x_2 which satisfy both equations. These values can be found either by drawing the graph of both equations on the same axes as illustrated in Fig. 1.3, or can be found by solving the two equations simultaneously as in Example 1.1. These two values are termed the roots of the equations.

Example 1.1
Find values for x_1 and x_2 which satisfy the pair of equations:

$$2x_1 - x_2 = 2 \tag{1}$$
$$6x_1 - 2x_2 = 9 \tag{2}$$

Step 1 Eliminate $6x_1$ from (2)

To do this a suitable multiplier for (1) is found such that on addition of the two equations, the x_1 term disappears. i.e. We divide the coefficient of the term we are eliminating by the value of the coefficient of the same term in the other equation, and reverse the sign. In this example the multiplier is $-6/2 = -3$

$$2x_1 - x_2 = 2 \tag{1}$$

multiply by -3

$$-6x_1 + 3x_2 = -6 \tag{3}$$
add $\quad 6x_1 - 2x_2 = 9 \tag{2}$

$$x_2 = 3$$

Step 2   Substitute for x_2 in either of the original equations to find x_1.
   Using (1):

$$2x_1 - 3 = 2$$
$$2x_1 = 2 + 3$$
$$x_1 = 2.5$$

A graphical solution for these equations is illustrated as Fig. 1.3.

By being logical in the way the multipliers are calculated and being systematic in the choice of terms which are eliminated, this method can be extended to solve *any* set of simultaneous linear equations (SLE) the general form of which is;

$$a_{1,1}x_1 + a_{1,2}x_2 + a_{1,3}x_3 + \cdots + a_{1,n}x_n = z_1$$
$$a_{2,1}x_1 + a_{2,2}x_2 + a_{2,3}x_3 + \cdots + a_{2,n}x_n = z_2$$

$$a_{m,1}x_1 + a_{m,2}x_2 + a_{m,3}x_3 + \cdots + a_{m,n}x_n = z_m$$

where $a_{1,1}$ to $a_{m,n}$ are given coefficients of unknown roots, x_1 to x_n z_1 to z_m are given values for the right-hand side (RHS) of the equations.

There will be a unique solution provided that $n = m$, and that the m equations are **linearly independent**. As we shall show later, the equation

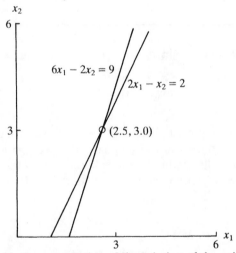

Figure 1.3 Graphical representation of the solution of the pair of simultaneous linear equations, used as Example 1.1:

$$2x_1 - x_2 = 2$$
$$6x_1 - 2x_2 = 9$$

The point of intersection (2.5, 3), gives the roots of the two equations.

will be linearly independent provided that the determinant of the matrix formed by the coefficients of the x_2 terms is not zero.

Terms can be eliminated in two ways. They can be reduced one stage at a time so that the only values left are in the **principal diagonal** i.e. positions in the equation set where both subscripts of the coefficients and that of the corresponding x-term are equal:

$$
\begin{aligned}
a_{1,1}x_1 \qquad\qquad\qquad\qquad &= z_1 \\
a_{2,2}x_2 \qquad\qquad\qquad &= z_2 \\
a_{3,3}x_3 \qquad\qquad &= z_3 \\
\cdot \qquad\quad &\;\; \cdot \\
\cdot \qquad\quad &\;\; \cdot \\
\cdot \qquad\quad &\;\; \cdot \\
a_{m,n}x_n &= z_m
\end{aligned}
$$

This is the **Gauss Jordan method**, also known as **reduced row echelon form**. Alternatively the equation set can be reduced to an **upper triangle of terms**;

$$
\begin{aligned}
a_{1,1}x_1 + a_{1,2}x_2 + a_{1,3}x_3 + \cdots + a_{1,n}x_n &= z_1 \\
a_{2,2}x_2 + a_{2,3}x_3 + \cdots + a_{2,n}x_n &= z_2 \\
a_{3,3}x_3 + \cdots + a_{3,n}x_n &= z_3 \\
\cdot \qquad\quad &\;\; \cdot \\
\cdot \qquad\quad &\;\; \cdot \\
a_{m,n}x_n &= z_m
\end{aligned}
$$

and finally, knowing the coefficient $a_{m,n}$, the root x_n can be found and the equation set can be solved to give the other roots, x_{n-1} to x_1. This method is known as **Gaussian elimination**, the equation set (as given above), is said to be in **row echelon form**. This method is more efficient in terms of the number of operations required to solve the equations, than for the Gauss Jordan method.

1.1.1 Geological examples

To introduce the application of simultaneous linear equations in geological studies a number of simple examples can be considered at this stage.

Example 1.2

A 'mixing problem' can often be tackled successfully using SLE, providing there are suitable quantifiable characteristics. Let us suppose we observe a student field-trip and students arrive singly either in their own cars or on their motor-cycles. A count of heads shows that there are z_1 students present and it is established that there are z_2 wheels on the student vehicles in the car park. As cars have 4 wheels (excluding spares!) and motor-cycles have 2 wheels, then the number of students who travelled by car x_1 and the number of students who travelled on their motor-cycles x_2, will be given by the solution of the equations:

$$x_1 + x_2 = z_1$$
$$4x_1 + 2x_2 = z_2$$

More complicated mixing situations will be dealt with later in the text. Of particular importance are applications in the fields of petrology and well–log analysis.

In the field of palaeobiometrics, the equation most frequently used to describe changes in size during growth is the linear equation, written in the form:

$$x_2 = ax_1 + c$$

where x_1 and x_2 are two measurable characteristics and the slope a and intercept c are estimated from a graphical plot of the data or are calculated using the equations developed in the next example. In some closely related invertebrates, similarities in growth pattern cause difficulties when separating individuals of different species. In this situation it can be useful to discover when two closely allied species will be the same size. The problem can be solved provided estimates of the constants of the growth equations for the species being studied are available, Fig. 1.4.

If the constants relating the width of the shells W with their lengths L of two species are a_1,c_1 and a_2,c_2 respectively then the two equations will be;

$$W = a_1L + c_1 \quad \text{(for species 1)}$$
$$W = a_2L + c_2 \quad \text{(for species 2)}$$

which can be rewritten as:

$$W - a_1L = c_1$$
$$W - a_2L = c_2$$

Putting the numerical values of a_1, c_1 and a_2, c_2 into these equations, their solution will give:

W – the measurement when the two species will be of equal width; and

L – the measurement when they will be of equal length;

both W and L being in the same units as the original measurements.

Since linearly correlated data occurs in many situations in geological data processing, it is convenient to consider a practical method for calculating the values for the slope and intercept, other than by graphical estimation. Let us suppose that a number of pairs of measurements x_1 and x_2 have been made on some objects of interest and a graph showing a plot of the data suggests that there may be a linear relationship. That is:

$$x_2 = ax_1 + c$$

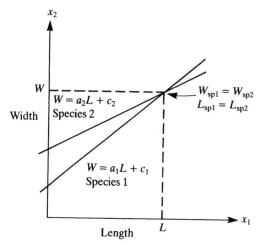

Figure 1.4 Illustration of the application of the solution of SLE to the problem of finding when the widths W and the lengths L of two closely related invertebrate species, are equal,
i.e. $W_{sp1} = W_{sp2}$ and $L_{sp1} = L_{sp2}$.

then using the data available, the numerical values for the slope and intercept of the fitted straight line may be calculated using the **least squares criterion**. This means that if one of the measured variables is taken as being fixed and the other is then calculated on the basis of the fitted line as in Fig. 1.5, the sums of squares of the differences between the observed and calculated values is minimized. That is:

$$\Sigma (x_{2obs} - x_{2calc})^2 = \text{minimum}$$

(The symbol Σ – sigma is used to imply summation.)

and $$x_{2calc} = ax_1 + c$$

Thus for any measured value of x_1, the difference between x_{2obs} and x_{2calc} will be given by:

$$m_i = \text{difference} = x_{2obs} - (ax_1 + c)$$

Now, if the sum of the squared deviations is given by $F(a,c)$ then:

$$F(a,c) = \Sigma (x_{2obs} - ax_1 - c)^2$$

If $F(a,c)$ is to be minimized as stated earlier, then the partial differentials, should be equal to zero:

$$\frac{\delta F}{\delta a} = \frac{\delta F}{\delta c} = 0$$

using values of x_{2obs}, the partial derivatives are:

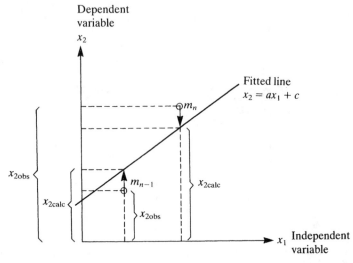

Figure 1.5 Graphical illustration of the least squares criterion, for finding the constants of the linear equation, given pairs of measurements made on a number of related objects. The fitted line $x_2 = ax_1 + c$ minimizes $\Sigma\ (x_{2\text{obs}} - x_{2\text{calc}})^2$.

$$\frac{\delta F}{\delta a} = \Sigma\ 2(x_2 - ax_1 - c)(-x_1) = 0$$

$$\frac{\delta F}{\delta c} = \Sigma\ 2(x_2 - ax_1 - c)(-1) = 0$$

which after multiplying out and taking into account n, the number of pairs of measurements, the equations which allow calculation of the unknowns (slope and intercept), are:

$$a\Sigma x_1^2 + c\Sigma x_1 = \Sigma x_1 x_2$$
$$a\Sigma x_1 + c_n = \Sigma x_2$$

which is a pair of SLE where a and c are the roots of the equations. This argument can be extended to give sets of equations to solve the same problem in three dimensions (Chapter 7).

It should be noted that when the equations were derived, the variable x_1 was arbitrarily chosen as the independent variable, hence it follows that the calculated value for x_2 based on that decision is the dependent variable. In certain situations it would have been proper to have chosen x_2 as the independent variable and then to have based the equations on $x_{1\text{obs}}$ and $x_{1\text{calc}}$. This would give rise to a second and different set of equations for the calculation of a and c. In many practical situations it is clear which is the dependent variable, in which case the correct set of equations must be used. When it is not clear, either variable can be arbitrarily chosen, or both can be used to calculate two equations relating x_1 and x_2. In this case it can be shown that the point of intersection of the two lines will be the means for the two variables measured.

Example 1.3

The relationship between the mean and the point of intersection is illustrated using an example from organic geochemistry. The data relating to the amount of the two gases methane and ethane held in fluid inclusions in fluorite (measured in volume parts per million per cc of mineral), is from samples collected from Cambokeels Mine, Co. Durham (Ferguson, 1991), and has been used to give equations for the regression of methane on ethane and vice versa. The two regression lines calculated after scaling methane, are plotted on the graph Fig. 1.6. The roots of the two equations:

$$1.0M - 1.253E = 0.707$$
$$-0.68M + 1.0E = 0.373$$

give the point of intersection (note that methane was scaled by $10E-2$):

$$M \text{ (methane)} = 7.9665$$
$$E \text{ (ethane)} = 5.79$$

The actual mean values for the data used are 791.0 (methane) and 5.75 (ethane). Students are left to verify this result for themselves (section 1.1.2, Q1.3).

As a final example in this section, a problem involving the balancing of chemical reactions, using an example from metamorphic petrology, illustrates the use of SLE in a straightforward manner.

Example 1.4

Consider the metamorphic reaction:

$$\text{grossular} + \text{quartz} = \text{wollastonite} + \text{anorthite}$$

This equation is written in terms of the reactants and products, without any indication of the number of molecules involved. The problem is to determine the numbers of molecules of the individual minerals which will balance the equation.

First of all we assign x_1, x_2, x_3 and x_4 as the number of molecules required, then we write the equation in terms of its chemical formulae:

$$x_1\, Ca_3Al_2Si_3O_{12} + x_2\, SiO_2 - x_3\, CaSiO_3 = x_4\, CaAl_2Si_2O_8$$

note that we have moved one product, wollastonite on to the left-hand side of the equation and obeying the algebraic rule, have reversed the sign. We can now expand the equation and rewrite the information it contains in terms of oxides and minerals:

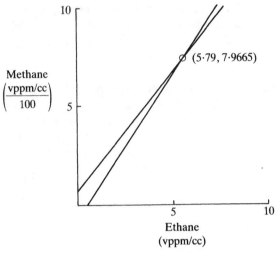

Figure 1.6 Graphical representation of the solution of two simultaneous linear equations:
$$M = 1.253E + 0.707$$
$$E = 0.68M + 0.377$$
where M = methane measured in vppm/cc * 10^{-2}
E = ethane measured in vppm/cc
The two equations were obtained by regressing methane against ethane and vice versa, as explained in the text (Example 1.3). The point of intersection (5.79,7.9665), gives the mean values for the data which were used to generate the regression equations.

grossular + quartz − wollastonite = anorthite

Al_2O_3	$1x_1$	$0x_2$	$0x_3$	$1x_4$
CaO	$3x_1$	$0x_2$	$-1x_3$	$1x_4$
SiO_2	$3x_1$	$1x_2$	$-1x_3$	$2x_4$

Which is a convenient way of expressing the information in the chemical formulae for the reaction, in view of the problem posed. This arrangement expresses the fact that x_1 moles of grossular, containing one mole of Al_2O_3, three moles of CaO and three moles of SiO_2, plus x_2 moles of quartz (equivalent to one mole SiO_2), less x_3 moles of wollastonite containing one mole of CaO and one mole of SiO_2, give x_4 moles anorthite which contain one mole Al_2O_3, one mole CaO and two moles SiO_2. Also if we make an assumption about the amount of anorthite x_4 involved in the reaction, then we can write an equation set for the data. Let us suppose that the mole proportion of anorthite is one, then the equation set will be:

$$1x_1 \qquad\qquad = 1 \qquad\qquad (1)$$
$$3x_1 \qquad - 1x_3 = 1 \qquad\qquad (2)$$
$$3x_1 + 1x_2 - 1x_3 = 2 \qquad\qquad (3)$$

It is now possible to solve the equations for the other three unknowns:
   Thus from (1) we can see that $x_1 = 1$ which allows solution of (2), whence $x_3 = 2$ and finally, from (3), $x_2 = 1$
   The reaction is, therefore:

$$1 \text{ grossular} + 1 \text{ quartz} = 2 \text{ wollastonite} + 1 \text{ anorthite}$$

It is important to note at this stage that since the sign for wollastonite was reversed when it was moved to the left-hand side of the equation, the value calculated for the mole proportion is positive. If the sign had not been reversed then the value calculated would have been negative indicating that it belonged on the other side of the equation. The student should satisfy himself that this is true and remember; if for some reason it is not certain whether a mineral is a reactant or a product, then the algebraic solution should differentiate between them with appropriate signs.

1.1.2 Student examples

Q1.1   Twenty students attend a field trip, travelling singly either by car or by motor cycle. If there are 74 wheels in the car park, how many cars and how many motor cycles were used?

Q1.2   In a study of the Carboniferous Rhynchonellid brachiopod *Pugnax pugnus* (Martin), Parkinson (1969) gives the following equations relating the growth of the length of the valves to the width (measurements in mms):

Treak cliff collection 3	$L = 0.644W + 2.13$
Thorpe cloud	$L = 0.777W - 0.1$
Bolland	$L = 0.690W + 0.8$

Find the points of intersection, for each possible combination of the three localities. From this draw up a table showing the measurements when the lengths and widths of the shells from each pair of localities are equal.

Q1.3   The two regression equations used in Example 1.3 are:

$$\text{Methane} = 1.253 \text{ Ethane} + 0.707$$
$$\text{Ethane} = 0.68 \text{ Methane} + 0.373$$

Find the means for the two gases (Methane value was scaled by $10E-2$).

Q1.4   The following data obtained from the analysis of sediment samples from Triassic argillaceous sediments from near Buraydah, Saudi Arabia, relate to the amount of lead and zinc (values in ppm) in the sediment: (42.7, 14.9); (85.8, 23.2); (57.0, 23.3); (43.0, 16.6); (10.0, 8.0) and (11.0, 8.0). Using lead as the dependent variable and assuming that there is a linear relationship between the metals, calculate the constants (to 2 sig. figs.) in the equation:

$$Pb = aZn + c$$

Q1.5   What are the molecular proportions involved in the reaction,

$$enstatite + periclase - FeO = forsterite$$

Assume that the proportion of forsterite is 1.0.

1.2 Methods of solution of simultaneous linear equations

1.2.1 Gaussian elimination

Returning to the problem of solving simultaneous linear equations we consider in more detail the method known as Gaussian elimination.

Example 1.5
The following set of equations describe the formation of the three minerals talc, chrysotile and sepiolite. We are given as data the Gibbs free energy (GFE) of formation of the three minerals. The problem is to find values for the GFE of the three compounds involved in the reaction. These are magnesium hydroxide, magnesium oxide and silicon dioxide.
   The equations are:

$$Mg(OH)_2 + 2MgO + 4SiO_2 = talc\ (-5523\ kjmol^{-1})$$
$$2Mg(OH)_2 + MgO + 2SiO_2 = chrysotile\ (-4038\ kjmol^{-1})$$
$$2Mg(OH)_2 + 3SiO_2 = sepiolite\ (-4269.8\ kjmol^{-1})$$

The first step is to rewrite these as simultaneous linear equations so that the roots x_1, x_2 and x_3 are the GFE required, the rows and columns are labelled for clarity at this stage:

	$Mg(OH)_2$		MgO		SiO_2		GFE
Talc	$1x_1$	$+$	$2x_2$	$+$	$4x_3$	$=$	-5523
Chrysotile	$2x_1$	$+$	$1x_2$	$+$	$2x_3$	$=$	-4038
Sepiolite	$2x_1$			$+$	$3x_3$	$=$	-4269.8

We solve the SLE using the Gaussian elimination

$$1x_1 + 2x_2 + 4x_3 = -5523 \qquad (1)$$
$$2x_1 + 1x_2 + 2x_3 = -4038 \qquad (2)$$
$$2x_1 \quad\ + 3x_3 = -4269.8 \qquad (3)$$

Step 1   Find multipliers for the x_1 coefficient (known as the **first pivot**) in (1), known as the **first pivotal equation**, which will enable us to eliminate the x_1 coefficient in (2) and (3).

Remember, the multiplier is found by dividing the coefficient of the x-term we wish to eliminate by the coefficient of the particular pivot (in this case the first pivot), and reverse the sign. The multipliers for this example are both -2.

$$1x_1 + 2x_2 + 4x_3 = -\ 5523 \tag{1}$$

multiply by
$$-2$$

gives
$$-2x_1 - 4x_2 - 8x_3 = +11046$$

add
$$2x_1 + 1x_2 + 2x_3 = -\ 4038 \tag{2}$$

gives
$$-\ 3x_2 - 6x_3 = +\ 7008 \tag{4}$$

similarly, using (3) we get:

$$-\ 4x_2 - 5x_3 = +\ 6776.2 \tag{5}$$

The equation set is now:

$$\begin{aligned} 1x_1 + 2x_2 + 4x_3 &= -5523 \tag{1}\\ -\ 3x_2 - 6x_3 &= +7008 \tag{4}\\ -\ 4x_2 - 5x_3 &= +6776.2 \tag{5} \end{aligned}$$

Note: Equation (1) is unchanged.

Step 2   Find a multiplier for the x_2 coefficient (second pivot) in (4), to eliminate the x_2 coefficient in (5). The multiplier is $-4/3$.

$$-3x_2 - 6x_3 = +7008 \tag{4}$$

multiply by
$$-4/3$$

gives
$$4x_2 + 8x_3 = -9344$$

add
$$-4x_2 - 5x_3 = +6776.2 \tag{5}$$

gives
$$3x_3 = -2567.8 \tag{6}$$

The equation set is now:

$$\begin{aligned} 1x_1 + 2x_2 + 4x_3 &= -5523 \tag{1}\\ -\ 3x_2 - 6x_3 &= +7008 \tag{4}\\ 3x_3 &= -2567.8 \tag{6} \end{aligned}$$

Note: Equations (1) and (4) have remained unchanged and (5) has changed as a result of Step 2 to give (6), i.e. the pivotal equations used to find the multipliers do not change.

Step 3   The roots can now be found by back calculation:
$$x_3 = \frac{-2567.8}{3}$$
$$= -855.93$$

$$x_2 = \frac{(-7008 - (-855.93 * 6.0))}{3}$$
$$= -624.13$$
$$x_1 = -5523 + (855.93 * 4.0) + (624.13 * 2.0)$$
$$= -851$$

Hence the Gibbs free energies required are:

$$G^O \ Mg(OH)_2 \ = \ - \ 851 \ kjmol^{-1}$$
$$G^O \ MgO \qquad = \ - \ 624.13 \ kjmol^{-1}$$
$$G^O \ SiO_2 \qquad = \ - \ 855.93 \ kjmol^{-1}$$

1.2.2 Student examples

Q1.6 Solve using Gaussian elimination:

$$1x + 1y - 1z = 1 \qquad (1)$$
$$8x + 3y - 6z = 5 \qquad (2)$$
$$-4x - 1y + 2z = 3 \qquad (3)$$

Q1.7 Solve using Gaussian elimination:

$$x_1 \ - \ x_2 \ + x_3 \ - \ 2x_4 = 8 \qquad (1)$$
$$2x_1 \ - \ x_2 \ + 2x_3 + x_4 \ = 5 \qquad (2)$$
$$-x_1 + x_2 \ + 2x_3 - x_4 \ = 4 \qquad (3)$$
$$x_1 \ + 2x_2 + 4x_3 + x_4 \ = 5 \qquad (4)$$

1.2.3 The Gauss–Seidel method

The method of solving equations just described is a direct method, and in most cases, will lead to an exact solution. An alternative to this would be an indirect numerical solution where each equation is rewritten in terms of one of the roots. These are then used by inserting an initial guess(es) for the other root(s) on the RHS of each of the equations and finding an estimate for the roots (x–terms on LHS) by calculation. By repeating the procedure and systematically updating the estimates as each new value is found, approximations to the true values of the unknowns can be found to any desired degree of accuracy. The method to be outlined is known as the **Gauss–Seidel iteration**.

Example 1.6
Solve the following set of equations using the Gauss–Seidel iteration:

$$4x_1 + 2x_2 = 1$$
$$x_1 + 3x_2 = 4$$

Step 1 Rewrite the two equations in terms of x_1 and x_2 respectively:

$$x_1 = \frac{(1 - 2x_2)}{4} \qquad (1)$$

$$x_2 = \frac{(4 - x_1)}{3} \tag{2}$$

Step 2   Commence the iteration by setting a value for x_2 and solving equation (1)

Let $x_2 = 1$

Then

$$x_1 = \frac{(1 - 2)}{4}$$

$$= -0.25$$

Step 3   Put the calculated value for x_1 into (2) and solve

Then

$$x_2 = \frac{(4 - (-0.25))}{3}$$

$$= 1.4167$$

Continue the iteration, repeating steps 2 and 3 using the current updated values for x_1 and x_2:

$$x_1 = \frac{(1 - 2(1.4167))}{4} \quad = -0.4583$$

$$x_2 = \frac{(4 - (-0.4583))}{3} = 1.4861$$

$$x_1 = \frac{(1 - 2(1.4861))}{4} = -0.49305$$

$$x_2 = \frac{(4 - (-0.49305))}{3} = 1.49768$$

This process can be continued, the difference between successive estimates becoming smaller as the calculated values approach the true value of the roots. In other words the solution will converge toward the true values, but may never reach it. Hence the iteration should be terminated when the difference between successive estimates is less than some predetermined level, which is related to the accuracy to which the values are required.

In the example used in this illustration the true values are:

$$x_1 = -0.5$$
$$x_2 = 1.5$$

The solution to this problem is shown graphically in Fig. 1.7, which demonstrates how successive iterations converge on the point of intersection.

The general set of linear equations can be solved by writing in the form:

$$x_1 = (z_1 - a_{1,2}x_2 - a_{1,3}x_3 - \ldots - a_{1,n}x_n)(1/a_{1,1})$$
$$x_2 = (z_2 - a_{2,1}x_1 - a_{2,3}x_3 - \ldots - a_{2,n}x_n)(1/a_{2,2})$$
$$\cdot \qquad \cdot$$
$$\cdot \qquad \cdot$$
$$\cdot \qquad \cdot$$
$$x_n = (z_m - a_{m,1}x_1 - a_{m,2}x_2 - \ldots - a_{m,m}x_m)(1/a_{m,n})$$

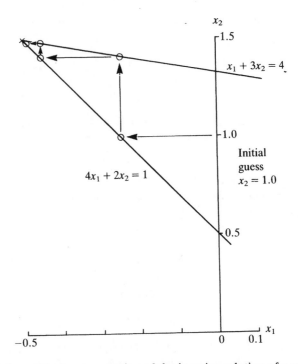

Figure 1.7 A graphical representation of the iterative solution of a pair of simultaneous linear equations. The graph shows the solution of the equations used in Example 1.6:

$$4x_1 + 2x_2 = 1$$
$$x_1 + 3x_2 = 4$$

Small circles represent the values (x_1, x_2) for each iteration starting with the initial guess $x_2 = 1.0$. Arrowed lines show the progress of the calcuation through the first five iterations.

In the example which has just been worked through each successive estimate x_i was used as it was calculated. In the general case all of the unknowns are set to some predetermined value, usually unity or zero and the updating done either as each new value is calculated – known as the method of **successive corrections** (Gauss–Seidel method), or at the end of each cycle of calculation. This is the method of **simultaneous corrections** (Jacobi method). Either method may be satisfactory although there are occasions when one method works rather than the other. For further details of this topic students should consult a text on numerical methods such as that of Mathews (1992). Occasionally, as with other iterative procedures, the solution may not converge as in the last example and students must be aware of this possibility. This is illustrated in the next example.

Example 1.7

Solve the equations given in Example 1.1 using the Gauss–Seidel iterative method. The equation set is:

$$2x_1 - x_2 = 2$$
$$6x_1 - 2x_2 = 9$$

where $x_1 = 2.5$ and $x_2 = 3.0$

Rewriting the equations in terms of x_1 and x_2:

$$x_1 = \frac{(2 + x_2)}{2}$$

$$x_2 = \frac{-(9 - 6x_1)}{2}$$

Using an initial guess: $x_2 = 1$ and tabulating the results for convenience:

Unknown	Iteration	1	3	5	10
x_1		1.5	0.25	−2.56	−35.94
x_2		0	−3.75	−12.19	−112.33

Clearly the successive estimates of the unknowns are increasing in value, and after a relatively small number of iterations the absolute values have become quite large. This example illustrates **divergence** which is due to the coefficients of the equations not being **diagonally dominant**. In other words, the largest coefficients of the x-terms are not in the principal diagonal (that is in the NW and SE positions) and as can be seen, the largest value is in the other diagonal, in the SW corner. Students should note that it is sometimes convenient to refer to coefficients or groups of coefficients in an equation set, by their geographic positions. Thus we can say:

In general a Gauss–Seidel iteration will diverge if the dominant coefficients are not in the principal diagonal.

In the example above, the equations can easily be rearranged so that they are diagonally dominant, giving:

$$6x_1 - 2x_2 = 9$$
$$2x_1 - x_2 = 2$$

The results obtained by solving the equations in this form are tabulated for convenience, using $x_2 = 0.0$ as the initial guess, then:

Unknown	Iteration	1	2	4	6	11	13	15
x_1		1.5	1.83	2.20	2.37	2.48	2.49	2.50
x_2		1.0	1.67	2.41	2.74	2.97	2.99	3.0

In this case the appearance of the coefficient value 6.0 for x_1 as the NW element of the coefficients, is sufficient to ensure that the iteration converges.

In conclusion it can be said that, provided that the coefficients of a set of simultaneous linear equations are diagonally dominant, then the use of an iterative method of solution has advantages. In particular if there are a large number of zero coefficients or if the matrix of coefficients is 'ill-conditioned' – a factor which will be considered in more detail later in Chapter 4.

1.2.4 Student examples

Q1.8 Using the iterative method outlined, solve the following equation sets.

(a) $4x_1 - 2x_2 = 1$ (b) $6x_1 + 2x_2 = 9$
     $x_1 - 3x_2 = 4$ $2x_1 + 1x_2 = 2$

Use no more than 5 iterations to illustrate that the solutions are convergent.

Q1.9 Show that there is no convergent solution to the Gibbs free energy example used earlier to illustrate the method of Gaussian elimination. Hint: scale the values for the RHS of the equation. Do no more than 4 iterations.

1.3 Problem equations

Not every set of linear equations has a unique solution. If two equations differ only by the amount of the constant term (i.e. the value on the RHS), then there will be no solution. Fig. 1.8a illustrates this, showing that the two lines representing the equations are parallel to one another. On the other hand, if both sides of a pair of equations are exact multiples, as for example:

$$4x_1 - 3x_2 = 10 \qquad\qquad (1)$$
$$12x_1 - 9x_2 = 30 \qquad\qquad (2)$$

where (2) is exactly three times (1), then the equations are not linearly independent and one equation is redundant. When plotted, the two curves representing the equations will be coincident as illustrated in Fig. 1.8b. Without further information being supplied it is not possible to find a unique solution and there will be an infinite number of possible solutions. In another situation, we have the following equations:

$$4x_1 - 3x_2 = 10 \qquad\qquad (1)$$
$$12x_1 - 9x_2 = 31 \qquad\qquad (2)$$

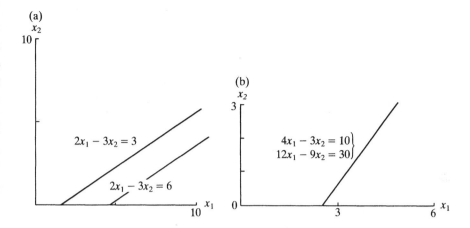

Figure 1.8 Examples of pairs of simultaneous linear equations which have no unique solution: (a) equations which differ only in the value of the constant; and (b) equations which are an exact multiple of one another.

where the LHS of (2) is an exact multiple of the LHS of (1), but the RHS is not. In this case the equations are not consistent and there will be no solution at all.

To sum up, provided that the equations in a set are all linearly independent i.e. one equation is not a multiple of another equation and that between any pair of equations the difference is not just in the value of the constant term, there will be a unique solution. If there is linear dependency between two or more equations there may be an infinite number of solutions or no solution at all. Further if an equation is almost a multiple of another equation the accuracy to which the calculation is performed and perhaps even the method used in the calculation will have a bearing on the solution. These matters will be dealt with more fully in Chapter 4.

These remarks are valid whether or not the number of roots equals the number of equations. However situations arise when these are not equal and there are more or fewer equations than roots. As these arise in a number of practical situations they will be dealt with now. Also as pointed out, when one equation is an exact multiple of another, then one equation is redundant and although the number of unknowns will remain the same, the number of equations will have been reduced by one.

1.3.1 Overdetermined equations

Overdetermined is the term used to describe the situation where there are more equations than unknowns.

Example 1.8
Find the roots x_1 and x_2 of the following set of three equations:

$$
\begin{array}{ll}
2x_1 - x_2 = 2 & \text{(1)} \\
6x_1 - 2x_2 = 9 & \text{(2)} \\
-3x_1 + 8x_2 = 13 & \text{(3)}
\end{array}
$$

In this example the equations cannot all be solved together and the best that can be done is to solve them pair-wise. The solution of (1) and (2) taken together we have already shown to be:

$$
\begin{aligned}
x_1 &= 2.5 \\
x_2 &= 3
\end{aligned}
$$

Solving (2) and (3) taken together:

$$
\begin{array}{ll}
6x_1 - 2x_2 = 9 & \text{(2)} \\
-3x_1 + 8x_2 = 13 & \text{(3)}
\end{array}
$$

The multiplier of (2) to eliminate the x_1-term from (3) is $\frac{3}{6} = \frac{1}{2}$

$$
\begin{array}{lll}
& 6x_1 - 2x_2 = 9 & \text{(2)} \\
\text{multiply by} & \quad\quad 0.5 & \\
\hline
& 3x_1 - x_2 = 4.5 & \\
\text{add} & -3x_1 + 8x_2 = 13 & \text{(3)} \\
\hline
& 7x_2 = 17.5 & \\
& x_2 = 2.5 & \\
\text{by substitution} & x_1 = 2.33 &
\end{array}
$$

Similarly, solving (1) and (3) taken together we get:

$$
\begin{aligned}
x_1 &= 2.2305 \\
x_2 &= 2.461
\end{aligned}
$$

Thus there is no single solution for x_1 and x_2 which is common to all three equations.

Example 1.9
Calculate the proportion of the end-members of the olivine series, forsterite and fayalite from analytical data relating to the proportions of the oxides of magnesium, iron and silicon in a sample using the following four equations (after Perry, 1967). In the equations x_1 represents the proportion of fayalite and x_2 the proportion of forsterite in the original sample. The equations are:

$$
\begin{array}{ll}
1x_1 + 1x_2 = 0.65 & \text{(1)} \\
2x_1 \quad\quad\quad = 0.26 & \text{(2)} \\
\quad\quad 2x_2 = 1.04 & \text{(3)} \\
4x_1 + 4x_2 = 2.60 & \text{(4)}
\end{array}
$$

Taking (2) and (3) the values for the unknowns can be calculated directly:

$$x_1 = 0.13$$
$$x_2 = 0.52$$

These solutions also satisfy equation pairs (1) and (2); (1) and (3); (2) and (4); (3) and (4). Note that in this example there are three sets of linearly independent equations and one pair ((1) and (4)), which are exact multiples of one another. We can conclude this example by calculating the proportions of the two end-members' minerals which are:

$$Fa_{20}, Fo_{80}$$

Example 1.10

In another similar example, based on the oxide analysis of a calcite/dolomite mixture, the equations relating the proportion of calcite x_1 to dolomite x_2 are: ·

$$x_1 + 0.5x_2 = 0.8318 \tag{1}$$
$$0.5x_2 = 0.1986 \tag{2}$$
$$x_1 + x_2 = 1.0304 \tag{3}$$
$$3x_1 + 3x_2 = 3.0912 \tag{4}$$

Here as in Example 1.9 there are three linearly independent equations (1), (2) and either (3) or (4) (these last two being multiples), which have a unique solution:

$$x_1 = 0.6332$$
$$x_2 = 0.3972$$

which gives the mineral mixture:

$$Ca_{61.4}, Do_{38.6}$$

1.3.2 Underdetermined equations

Underdetermined equations are systems where there are more unknowns than equations (the reverse of overdetermined systems).

Example 1.11

Consider the reaction:

$$x_1 \text{ grossular} + x_2 \text{ kyanite} + x_3 \text{ quartz} = x_4 \text{ anorthite}$$

which can be written in terms of the chemical formulae as:

$$x_1 \text{ Ca}_3\text{Al}_2\text{Si}_3\text{O}_{12} + x_2 \text{ Al}_2\text{SiO}_5 + x_3 \text{ SiO}_2 = x_4 \text{ CaAl}_2\text{Si}_2\text{O}_8$$

and written in terms of oxides:

	x_1	x_2	x_3	x_4
Al_2O_3	1	1	0	1
CaO	3	0	0	1
SiO_2	3	1	1	2

This problem is underdetermined as there are four roots and only three equations, and as we did in a similar example previously, we can solve for three roots if we assume a value for the fourth. To keep the arithmetic simple, let us assume that the molecular proportion of anorthite is 1, the equations are:

$$1x_1 + 1x_2 \qquad = 1 \qquad\qquad (1)$$
$$3x_1 \qquad\qquad = 1 \qquad\qquad (2)$$
$$3x_1 + 1x_2 + 1x_3 = 2 \qquad\qquad (3)$$

The three equations are linearly independent and can easily be solved to give:

$$x_1 = \tfrac{1}{3};\ x_2 = \tfrac{2}{3};\ x_3 = \tfrac{1}{3} \text{ when } x_4 = 1$$

Now, since by convention the stochiometric proportions of reactants and products in chemical equations are specified by integer values, we must now multiply through the lowest common denominator (LCD). In this case the LCD is 3, and the equations become:

$$1 \text{ grossular} + 2 \text{ kyanite} + 1 \text{ quartz} = 3 \text{ anorthite}$$

Example 1.12

As a further illustration we use a simple problem in **linear programming**.

Let us suppose that we are financial advisers to a small company producing standards for analytical work, of which a small section is concerned with the manufacture of cans of standard gas mixtures. The section can produce four different standards, which are mixtures of three light hydrocarbon gases held in an inert gas. Each of the standards has a different unit profit, which depends solely on the proportions of the three hydrocarbon gases in the mixture. Each day only certain volumes of the three hydrocarbon gases are available. The problem posed is how to maximize the profit from this section of the venture. The data is given in Table 1.1.

This information can be rewritten as a series of linear equations where the number of cans of the Standards A–D which can be produced on a daily basis is denoted by x_1–x_4 respectively. The number of cans produced will be subject to using no more than the maximum amount of gas available. These are the **constraint equations**:

Table 1.1 Data for Example 1.12

	Methane vols	Ethane vols	Propane vols	Unit profit £
Standard A	20	20	5	8
Standard B	20	10	0	11
Standard C	30	20	10	7
Standard D	10	5	5	9
Maximum amount gas (available per day)	900	740	200	

$$20x_1 + 20x_2 + 30x_3 + 10x_4 = 900 \qquad (1)$$
$$20x_1 + 10x_2 + 20x_3 + 5x_4 = 740 \qquad (2)$$
$$5x_1 \qquad + 10x_3 + 5x_4 = 200 \qquad (3)$$

and are subject to $x_1 - x_4$ being greater or equal to 0 (i.e. we cannot produce negative numbers of standards).

The function to be maximized relates profit to production (known as the **objective** function), is:

$$8x_1 + 11x_2 + 7x_3 + 9x_4 = z \qquad (4)$$

where z will be the profit per day.

The algebraic solution to this problem can be found by solving equations (1), (2) and (3) for the four roots x_1, x_2, x_3, x_4 and put the resulting values into (4) to enable us to calculate a value for z. Since there are more unknowns than equations, it is only possible to solve them by systematically setting one of the unknowns to zero, which will give four sets of values which can then be substituted in the objective function. The results are given in Table 1.2.

Table 1.2 Results for Example 1.12

	x_1	x_2	x_3	x_4
Solution 1	0	−4	58	−76
Solution 2	8	0	42	−52
Solution 3	29	10.5	0	11
Solution 4	25.3	8.7	7.3	0

Clearly the first two solutions are not feasible since they involve the production of negative numbers of standards. Substituting the remaining solutions in the objective function and ignoring decimals we get:

Solution 3   Profit = (8 * 29) + (11 * 10) + (7 * 0) + (9 * 11)
                            = £441

Solution 4   Profit $= (8 * 25) + (11 * 8) + (7 * 7) + (9 * 0)$
                                 $= £337$

Therefore to maximize profit the section needs to produce 29 units of Standard A, 10 units of Standard B and 11 units of Standard D. In our quest for profit the production of Standard C ceases.

It should be noted at this point that in the first example a solution to the problem was found because of the a priori information that there is a simple integer relationship between the mole proportions expressed in a chemical reaction. Hence the assumption that one product or reactant has a value of unity, allows for solution. In the second example the simple expedient of setting variables to zero was acceptable since the only condition made for the unknowns was that they should be equal to or greater than zero. Also there was no prerequisite that there must be a solution which satisfied all the variables at once. All that was required was that the function relating to profit be maximized.

1.3.3 Homogeneous equations

If in the general system of equations (section 1.1) the values $z_1, z_2, \ldots ,$ z_n are all zero, then the system of equations is said to be **homogeneous**. The system of equations used in Example 1.11, can be written in the form:

$$x_1 \text{ grossular} + x_2 \text{ kyanite} + x_3 \text{ quartz} - x_4 \text{ anorthite} = 0$$

which leads to the homogeneous system:

$$1x_1 + 1x_2 \qquad - 1x_4 = 0 \qquad\qquad (1)$$
$$3x_1 \qquad\qquad - 1x_4 = 0 \qquad\qquad (2)$$
$$3x_1 + 1x_2 + 1x_3 - 2x_4 = 0 \qquad\qquad (3)$$

As noted earlier, in the general linear system of equations where the number of equations equals the number of roots then there are three possibilities: a unique solution; no solution at all or an infinite number of solutions. The homogeneous system on the other hand will always have the **trivial** or **zero** solution where:

$$x_1 = x_2 = x_3 = \cdots = x_n = 0$$

with two possibilities: either the trivial solution is the only solution or there are an infinite number of solutions in addition to the trivial solution. These other solutions are termed **nontrivial** solutions.

Example 1.13
To illustrate the solution of homogeneous equations we will solve the equation set above using Gaussian elimination as described earlier, with the exception that we will ensure that the absolute value of the pivotal coefficients will be unity. The first step will be to write the equations (1)

to (3) in a suitable form bearing in mind that we would like the absolute value of the pivotal coefficients to be unity. (Students should note that there is no rule which states that the equations should be rewritten, as they can equally well be solved in their original form. The advantage of rewriting is simply in reducing the number of calculations.)

$$1x_3 - 2x_4 + 1x_2 + 3x_1 = 0 \tag{1}$$
$$- 1x_4 + 1x_2 + 1x_1 = 0 \tag{2}$$
$$- 1x_4 \quad\quad + 3x_1 = 0 \tag{3}$$

Since (1) and (2) are already in the form which we require all that is necessary is to eliminate the x_4 term in (3):

$$- 1x_4 + 1x_2 + 1x_1 = 0 \tag{2}$$

multiply by $\quad\quad\quad\quad -1$

$$+ 1x_4 - 1x_2 - 1x_1 = 0$$
$$\text{add} \quad\quad - 1x_4 \quad\quad + 3x_1 = 0 \tag{3}$$

$$- 1x_2 + 2x_1 = 0 \tag{5}$$

The equation set now becomes:

$$1x_3 - 2x_4 + 1x_2 + 3x_1 = 0 \tag{1}$$
$$- 1x_4 + 1x_2 + 1x_1 = 0 \tag{2}$$
$$- 1x_2 + 2x_1 = 0 \tag{5}$$

Hence we have the trivial solution $x_1 = x_2 = x_3 = x_4 = 0$ or the nontrivial solution given by x_1. Thus if $x_1 = 1$ we have:

$$1x_3 - 2x_4 + 1x_2 = -3$$
$$- 1x_4 + 1x_2 = -1$$
$$- 1x_2 = -2$$

which on back calculation gives:

$$x_2 = 2; x_4 = 3; x_3 = 1$$

Therefore the reaction is:

$$1 \text{ grossular} + 2 \text{ kyanite} + 1 \text{ quartz} = 3 \text{ anorthite}$$

Before continuing we shall consider one further example.

Example 1.14
Rewrite the equations representing the following reaction in homogeneous form and hence find the stochiometric proportions of the reactants and products.

$$x_1 \text{ fayalite} + x_2 \text{ enstatite} = x_3 \text{ forsterite} + x_4 \text{ ferrosilite}$$

which can be written as:

$$x_1 \, Fe_2SiO_4 + x_2 \, MgSiO_3 - x_3 Mg_2SiO_4 - x_4 \, FeSiO_3 = 0$$

giving the system of equations (row and column labels are included for clarity):

	Fy		En		Fo		Fr		
SiO_2	$1x_1$	$+$	$1x_2$	$-$	$1x_3$	$-$	$1x_4$	$= 0$	(1)
FeO	$2x_1$					$-$	$1x_4$	$= 0$	(2)
MgO			$1x_2$	$-$	$2x_3$			$= 0$	(3)

Step 1   Eliminate the x_1-term from (2)

$$1x_1 + 1x_2 - 1x_3 - 1x_4 = 0 \tag{1}$$

multiply by -2

$$\begin{array}{l} -2x_1 - 2x_2 + 2x_3 + 2x_4 = 0 \\ 2x_1 \qquad\qquad\quad - 1x_4 = 0 \end{array} \tag{2}$$

add

$$- 2x_2 + 2x_3 + 1x_4 = 0 \tag{4}$$

Step 2   Divide (4) by 2 to drive the second pivot to unity:

$$- 1x_2 + 1x_3 + \tfrac{1}{2}x_4 = 0 \tag{5}$$

add $$1x_2 - 2x_3 \qquad\quad = 0 \tag{3}$$

$$- 1x_3 + \tfrac{1}{2}x_4 = 0 \tag{6}$$

The equation set is now:

$$\begin{array}{l} 1x_1 + 1x_2 - 1x_3 - 1x_4 = 0 \\ \quad - 1x_2 + 1x_3 + \tfrac{1}{2}x_4 = 0 \\ \qquad\quad - 1x_3 + \tfrac{1}{2}x_4 = 0 \end{array} \qquad \begin{array}{l} (1) \\ (5) \\ (6) \end{array}$$

Hence we have the trivial solution $x_1 = x_2 = x_3 = x_4 = 0$ or nontrivial solutions given by x_4. Thus if $x_4 = 1$ we have:

$$\begin{array}{l} 1x_1 + 1x_2 - 1x_3 = 1 \\ \quad - 1x_2 + 1x_3 = -\tfrac{1}{2} \\ \qquad\quad - 1x_3 = -\tfrac{1}{2} \end{array}$$

which on back calculation gives: $x_3 = \tfrac{1}{2}$, $x_2 = 1$, $x_1 = \tfrac{1}{2}$ clearing fractions, the reaction is:

$$1 \text{ fayalite} + 2 \text{ enstatite} = 1 \text{ forsterite} + 2 \text{ ferrosilite}$$

These examples, where there is an infinite number of nontrivial solutions illustrates the general rule for homogeneous systems, that if there are more roots than equations there will be the trivial or zero solution plus an infinite number of nontrivial solutions, whereas if the number of roots equals the number of equations and no equation is redundant, then only the trivial or zero solution is possible.

1.3.4 Other geological applications

In crystallography If we wish to find the zone axis symbol $[U, V, W]$ for a zone containing two generalized crystal faces whose Miller indices are $(h_1 \ k_1 \ l_1)$ and $(h_2 \ k_2 \ l_2)$, the problem may be solved using the homogeneous equations:

$$h_1 U + k_1 V + l_1 W = 0$$
$$h_2 U + k_2 V + l_2 W = 0$$

which are underdetermined and have the trivial solution $U = V = W = 0$ and an infinite number of nontrivial solutions. The equations can be solved since by definition U, V and W are integer numbers with no other common factor than unity.

Example 1.15

Find the zone axis symbol for the zone containing the faces (2 1 0) and (0 1 1).

The equations are:

$$2U + 1V \qquad\quad = 0 \qquad\qquad (1)$$
$$1V + 1W = 0 \qquad\qquad (2)$$

From (2) if $W = 1$ then $V = -1$
From (1) if $V = -1$ then $U = \frac{1}{2}$
Getting rid of fractions gives $U = 1$, $V = -2$ and $W = 2$
Thus $[1, -2, 2]$ is the zone axis symbol of the zone containing the faces (2 1 0) and (0 1 1).

Note: In this example as in other crystalographic problems which follow, we have used the negative sign rather than the conventional bar sign. Also notice the convention that the zone symbol $[U, V, W]$ is enclosed in square braces. This is to indicate that a line and not a plane is referred to. (The Miller index refers to a plane and can be enclosed in ordinary braces and without commas between the several elements.) For further information on these important topics see Windle (1977) or Battey (1981).

In petrology The application of Gibbs phase rule to metamorphic reactions is valid only if there are no compositional degeneracies (i.e. linear dependencies) among the mineral assemblages under consideration. Unfortunately there are occasions when it is not easy to determine by inspection whether a set of mineral reactions are all linearly independent. In such situations it is useful to have a test to establish linear independence, particularly where there are more end members than reactions.

Example 1.16

Are the following equilibrium relations all linearly independent?

$$\text{quartz} + \text{forsterite} = 2 \text{ enstatite}$$

$$\text{quartz} + \text{fayalite} = 2 \text{ ferrosilite}$$

$$\text{fayalite} + 2 \text{ enstatite} = \text{forsterite} + 2 \text{ ferrosilite}$$

substituting x_1 for quartz, x_2 for fayalite, x_3 for forsterite, x_4 for ferrosilite and x_5 for enstatite, and writing as homogeneous equations:

$$
\begin{aligned}
1x_1 \qquad\; + 1x_3 \qquad\quad - 2x_5 &= 0 \qquad (1)\\
1x_1 + 1x_2 \qquad\quad - 2x_4 \qquad\quad\; &= 0 \qquad (2)\\
1x_2 - 1x_3 - 2x_4 + 2x_5 &= 0 \qquad (3)
\end{aligned}
$$

We solve the equations:

$$1x_1 \qquad + 1x_3 \qquad\quad - 2x_5 = 0 \qquad (1)$$

multiply by $\qquad\qquad\qquad\qquad\qquad\quad -1$

$$
\begin{aligned}
-1x_1 \qquad\; - 1x_3 \qquad\quad + 2x_5 &= 0\\
1x_1 + 1x_2 \qquad\quad - 2x_4 \qquad\quad\; &= 0 \qquad (2)
\end{aligned}
$$

add

$$1x_2 - 1x_3 - 2x_4 + 2x_5 = 0 \qquad (4)$$

The equation set is now:

$$
\begin{aligned}
1x_1 \qquad + 1x_3 \qquad\quad - 2x_5 &= 0 \qquad (1)\\
1x_2 - 1x_3 - 2x_4 + 2x_5 &= 0 \qquad (3)\\
1x_2 - 1x_3 - 2x_4 + 2x_5 &= 0 \qquad (4)
\end{aligned}
$$

and as can be seen, one of the last pair of equations is redundant. This demonstrates that there is linear dependency among the original reactions, therefore we cannot apply the Phase rule to the system. As will be demonstrated later, there are other methods of testing for linear dependency in similar situations, and we shall return to this topic again.

1.3.5 Student examples

Q1.10 For the reaction:

$$\text{grossular} + \text{quartz} = \text{wollastonite} + \text{anorthite}$$

find the number of moles of each reactant and product by writing and solving the homogeneous equation set.

Q1.11 Four possible end-members of the orthochlorites are:

amesite	$Mg_4\,Al_2\,(Si_2\,Al_2)\,O_{10}\,(OH)_8$
antigorite	$Mg_6\,Si_4\,O_{10}\,(OH)_8$
ferroantigorite	$Fe_6\,Si_4\,O_{10}\,(OH)_8$
daphnite	$Fe_4\,Al_2\,(Si_2\,Al_2)\,O_{10}\,(OH)_8$

Are they linearly independent?

Hint: Use the elements Si, Al, Fe, and Mg as the unknowns $x_1 - x_4$ (column labels), minerals by rows.

Q1.12   The following equations relate to the possible reaction:

enstatite + hedenburgite = ferrosilite + diopside

SiO_2	$2x_1 - 2x_3 - 2x_4 + 2x_2 = 0$
MgO	$2x_1 \qquad - 1x_4 \qquad = 0$
FeO	$- 2x_3 \qquad + 1x_2 = 0$
CaO	$- 1x_4 + 1x_2 = 0$

Are the four components linearly independent?

Q1.13   Find the zone axis symbol of the zone containing the crystal faces:

(a) $(0\ 1\ 1)$ and $(-1\ -1\ 1)$
(b) $(1\ -1\ 0)$ and $(1\ 2\ 3)$

Q1.14   Find the stochiometric proportions for the following reaction:

quartz + muscovite + kyanite + chlorite + garnet + water = biotite

The chlorite is: $Mg_7\ Fe_2\ Al_6\ Si_5\ O_{20}\ (OH)_{16}$

Hint: use $AlO_{3/2}$ and $KO_{1/2}$ for aluminium and potassium oxides. Set the mole proportion for biotite equal to unity.

2

Scalars, vectors and matrices

The basic definitions and operations in matrix algebra will be dealt with in this chapter. For students who have not had previous experience of this algebra, it is important to note that many aspects of traditional algebra have their analogues in matrix form. Students should take particular note of the shorthand notation used, which is similar to that used in ordinary algebra, but frequently has a different meaning in terms of arithmetic operations.

2.1 Scalars and vectors – some basic definitions

2.1.1 Scalars

A scalar is a single number which can be real, integer or complex, positive or negative. It obeys the rules of ordinary arithmetic with respect to operations. In the context of matrix operations scalars are normally denoted by a Greek letter in order to differentiate them from vectors and matrices, when using algebraic shorthand to express mathematical operations.

2.1.2 Vectors

A vector is an n–component ordered set of numbers, either in row or column form. For example:

$$[x_1 \; x_2 \; \ldots \; x_n] \text{ is a } \textbf{row} \text{ vector}$$

and
$$\begin{bmatrix} x_1 \\ x_2 \\ . \\ . \\ . \\ x_n \end{bmatrix} \text{ is the corresponding } \textbf{column} \text{ vector}$$

Order is an important and essential part of the definition and two vectors with the same components but written in different orders are not the same, for example the vector:

$$\mathbf{b} = \begin{bmatrix} 0.65 \\ 0.26 \\ 1.04 \end{bmatrix} \text{ represents the chemical analysis}$$

of an olivine expressed as moles of the three component oxides: SiO_2, FeO and MgO in that order (Example 1.9).

A vector is denoted by a lower case letter in bold face, individual elements of the vector are represented by the letter with the appropriate index indicating their position. In the example just used, the vector is **b** and the element representing the number of moles of MgO in the analysis is b_3. In another example calculated from the analysis of a chlorite (original analytical data from Deer, Howie and Zussman, 1962, Table 25, Analysis 14), the vector **w** contains the number of ions in the mineral analysed. In this case $i = 6$ and the vector is:

$$\begin{matrix} Si \\ Al \\ Fe^{++} \\ Mg \\ H_2 \\ O \end{matrix} \begin{bmatrix} 0.444 \\ 0.494 \\ 0.121 \\ 0.669 \\ 0.649 \\ 3.068 \end{bmatrix} = w$$

The row labels (elements) are given for reference.

In vector arithmetic the following rules apply:
Let **x**, **y** and **z** be n vectors and α and β are scalars, then:

1. If a scalar β is added to a vector **x** then β is added to each individual value:

$$\beta + \begin{bmatrix} x_1 \\ x_2 \\ . \\ . \\ x_n \end{bmatrix} = \begin{bmatrix} x_1 + \beta \\ x_2 + \beta \\ . \\ . \\ x_n + \beta \end{bmatrix}$$

from this it follows that if $\beta = 0$, then:

$$x + \beta = x$$

2. If a vector **x** is multiplied (or divided) by a scalar α, then each element is multiplied (or divided) by α.

$$\alpha * \begin{bmatrix} x_1 \\ x_2 \\ . \\ . \\ x_n \end{bmatrix} = \begin{bmatrix} x_1 * \alpha \\ x_2 * \alpha \\ . \\ . \\ x_n * \alpha \end{bmatrix}$$

from this it also follows that if $\alpha = 0$, then

$$\alpha * x = 0$$

where **0** represents the zero vector.

3. Vectors can be added or subtracted provided that they are of the same size, by adding or subtracting corresponding elements. Vector addition is **cummutative**, i.e.:

$$x + y = y + x$$

4. The following results also hold:
    (a) $(x + y) + z = x + (y + z)$ – **associative law**
    (b) $\beta(x + y) = \beta x + \beta y$ – **distributive law**
    (c) $(\alpha + \beta)x = \alpha x + \beta x$
    (d) $(\alpha\beta)x = \alpha(\beta x) = \beta(\alpha x)$

In an earlier example a vector was given which expressed the analysis of a chlorite in terms of the number of ions. An alternative would be to state the vector in terms of the number of oxygen atoms required by the formulae. This would involve the scalar multiplication of the vector by a factor calculated as the number of oxygens required by the mineral formulae, divided by the number of oxygens deduced from the analysis.

Example 2.1
Recast the vector w given earlier in terms of the 14 oxygens required by the chlorite mineral formulae. The operation is:

$$\begin{bmatrix} 0.444 \\ 0.494 \\ 0.121 \\ 0.669 \\ 0.649 \\ 3.068 \end{bmatrix} * \frac{14}{3.068} = \begin{bmatrix} 2.026 \\ 2.254 \\ 0.552 \\ 3.053 \\ 2.962 \\ 14.00 \end{bmatrix}$$

The order of the elements is the same as that given earlier, and from this we conclude that the formulae for the particular chlorite mineral analysed is:

$$Si_2 (Al, Fe)_{2.81} Mg_3 H_6 O_{14}$$

2.1.3 The product of two vectors

Two vector products are defined, but for the moment we shall only define one, that is the **scalar** or **dot product** of two vectors.

Let $x = (x_1 \, x_2 \, \ldots \, x_n)$ and $y = (y_1 \, y_2 \, \ldots \, y_n)$

$$\text{or } x = \begin{bmatrix} x_1 \\ x_2 \\ . \\ . \\ x_n \end{bmatrix} \text{ and } y = \begin{bmatrix} y_1 \\ y_2 \\ . \\ . \\ y_n \end{bmatrix}$$

then $$\mathbf{x} \cdot \mathbf{y} = x_1 y_1 + x_2 y_2 + \ldots + x_n y_n$$

Notice that in order to obtain the product, the number of elements in the two vectors must be equal. Note also that the result of the operation is a scalar. The rule applies whether we are dealing with row or column vectors. The scalar or dot product of a row and column vector is calculated similarly:

$$(x_1 \ x_2 \ \ldots \ x_n) \cdot \begin{bmatrix} y_1 \\ y_2 \\ \cdot \\ \cdot \\ y_n \end{bmatrix} = x_1 y_1 + x_2 y_2 + \ldots + x_n y_n$$

Also note the use of the 'dot' to signify this product rather than the usual multiplication sign.

In section 1.3.2, Example 1.12, two feasible solutions to a problem in linear programming were calculated. These related to the numbers of cans of gas standards which could be produced on a daily basis, based on data relating to the amount of raw material (gases) available and the amounts the different gases needed for each of four standards. To calculate the profit relating to each feasible solution, the unknowns $x_1 - x_4$ (numbers of cans of each of the standards) were multiplied by the objective function and summed, which is the dot products of the vectors of the feasible solutions and the objective function.

Example 2.2

Calculate the profit for the two feasible solutions to Example 1.12, using the given objective function. The data is given in Table 2.1.

Table 2.1 Data for Example 2.2

	x_1	x_2	x_3	x_4
Feasible solution (3)	29	10	0	11
Feasible solution (4)	25	8	7	0
Objective function	8	11	7	9

The operations are:

$$\text{PROFIT (1)} = \begin{bmatrix} 29 \\ 10 \\ 0 \\ 11 \end{bmatrix} \cdot \begin{bmatrix} 8 \\ 11 \\ 7 \\ 9 \end{bmatrix} \qquad \text{PROFIT (2)} = \begin{bmatrix} 25 \\ 8 \\ 7 \\ 0 \end{bmatrix} \cdot \begin{bmatrix} 8 \\ 11 \\ 7 \\ 9 \end{bmatrix}$$

Profit (1) = (29 * 8) + (10 * 11) + (0 * 7) + (11 * 9) = £441.00
Profit (2) = (25 * 8) + (8 * 11) + (7 * 7) + (0 * 9) = £337.00

If **x**, **y** and **z** are n–vectors and α is a scalar, then:

(a) $\mathbf{x} \cdot \mathbf{0} = 0$ (**0** is the zero vector)

(b) $\mathbf{x} \cdot \mathbf{y} = \mathbf{y} \cdot \mathbf{x}$ – **commutative law**

(c) $\mathbf{x} \cdot (\mathbf{y} + \mathbf{z}) = \mathbf{x} \cdot \mathbf{y} + \mathbf{x} \cdot \mathbf{z}$ – **distributive law**

(d) $(\alpha * \mathbf{x}) \cdot \mathbf{y} = \alpha(\mathbf{x} \cdot \mathbf{y})$

2.1.4 Student examples

Q2.1   Multiply the following vectors by the given constants:

(a)   (2 5 6 8 9) (0.58)   (b)   (0 −1.5 2.3 4) (B)
(c)   (3 2 5 9) (2.5)   (d)   (9 2 −3 −4) (−3)

Q2.2   Find the dot product of the following vectors:
(a)   (2 3 1) (3 4 2)   (b)   (2 2 1) (3 4 2)
(c)   (3 7 4 2) (2 −1 3 2)   (d)   (2 −1 0 2) (−2 0 6 1)

Q2.3   Using the two vectors in (c) and (d) in Q2.2 above, show that the product (**a** . **b**) is equal to the product (**b** . **a**)

Q2.4   Using the three vectors below, show that the Distributive Law

$$\mathbf{x} \cdot (\mathbf{y} + \mathbf{z}) = (\mathbf{x} \cdot \mathbf{y}) + (\mathbf{x} \cdot \mathbf{z})$$

holds. The vectors are: $\mathbf{x} = (1\ 3\ 4\ 6)$
$$\mathbf{y} = (-1\ 3\ 9\ 5)$$
$$\mathbf{z} = (-3\ 2\ -4\ 7)$$

2.2 Matrices – some basic definitions

An $m \times n$ matrix **A** is a rectangular array of numbers arranged in m rows and n columns:

$$\mathbf{A} = \begin{bmatrix} a_{1,1} & a_{1,2} & \cdots & a_{1,j} & \cdots & a_{1n} \\ a_{2,1} & a_{2,2} & \cdots & a_{2,j} & \cdots & a_{2n} \\ & & & \vdots & & \\ a_{i,1} & a_{i,2} & \cdots & a_{i,j} & \cdots & a_{i,n} \\ & & & \vdots & & \\ a_{m,1} & a_{m,2} & \cdots & a_{m,j} & \cdots & a_{m,n} \end{bmatrix}$$

In practice m and n are not necessarily equal. If they are equal then the matrix is called a **square matrix**. The matrix is normally denoted by an upper case letter such as **A** in the example above, individual elements being indicated by a lower case letter with the appropriate subscripts indicating their position within the array. Sometimes the matrix **A** can be written as $\mathbf{A} = (a_{i,j})$. It should be noted that position within a matrix has meaning and although whole rows or columns can be changed over, the position of individual elements cannot be changed.

Example 2.3

Write the reactants of the following metamorphic reaction, as a matrix with the minerals as the column labels and the oxides as the row labels:

$$\text{grossular} + \text{kyanite} + \text{quartz} = \text{product}$$

The chemical formulae of the minerals are:

grossular – $Al_2\,Ca\,Si\,O_6$
kyanite – $Al_2\,Si\,O_5$
quartz – SiO_2

Therefore the matrix will be:

	grossular	kyanite	quartz
Al_2O_3	1	1	0
CaO	3	0	0
SiO_2	3	1	1

which is square matrix with three rows and columns:

$$m = n = 3$$

where the nth column tells us the number of moles of the mth oxide, thus element 3, 2 tells us that there is 1 mole of silica in the mineral kyanite.

Example 2.4

Write the mineral/oxide matrix for the two members of the hornblende series barkevite and glaucophane:

	barkevite	glaucophane
SiO_2	7	8
Al_2O_3	1	1
Fe_2O_3	0.5	0
FeO	1	0
MnO	0.5	0
MgO	0.5	3
CaO	2	0
Na_2O	0.5	1
H_2O	1	0

This matrix is not square and has 9 rows and 2 columns:

$$m = 9 \text{ and } n = 2$$

A column or row vector can be considered as a special matrix with 1 column (or row) and a specified number of rows (or columns).

Matrix notation provides a compact way of writing a set of SLE. The general set can be written as:

$$\begin{bmatrix} a_{1,1} & a_{1,2} & a_{1,3} & \cdots & a_{1,n} \\ a_{2,1} & a_{2,2} & a_{2,3} & \cdots & a_{2,n} \\ \cdot & \cdot & \cdot & & \cdot \\ \cdot & \cdot & \cdot & & \cdot \\ a_{m,1} & a_{m,2} & a_{m,3} & \cdots & a_{m,n} \end{bmatrix} \cdot \begin{bmatrix} x_1 \\ x_2 \\ \cdot \\ \cdot \\ x_m \end{bmatrix} = \begin{bmatrix} z_1 \\ z_2 \\ \cdot \\ \cdot \\ z_m \end{bmatrix}$$

or alternatively:

$$\mathbf{A} \cdot \mathbf{x} = \mathbf{z}$$

The **zero matrix** is a matrix of size $m \times n$ whose components are all equal to zero.

The **identity matrix** is a square matrix in which the values in the **principal diagonal** are all unity, the remainder are zero, i.e.

all elements i, j where $i = j$ are unity;

all others $i \neq j$ are zero.

The identity matrix is usually denoted by the symbol I.

Two matrices are said to be equal if they are of the same size and the numerical values of corresponding elements are equal.

The **transpose** of a matrix is the matrix written with its rows as columns and is denoted by:

$$\mathbf{A}^{\mathrm{T}} \text{ or } \mathbf{A}'$$

Example 2.5
Write the matrix A as its transpose:

$$\mathbf{A} = \begin{bmatrix} 2 & 3 & 1 \\ 4 & 3 & 2 \\ 1 & 1 & 1 \end{bmatrix} \quad \mathbf{A}^{\mathrm{T}} = \begin{bmatrix} 2 & 4 & 1 \\ 3 & 3 & 1 \\ 1 & 2 & 1 \end{bmatrix}$$

If a matrix and its transpose are equal, then the matrix is said to be **symmetrical**:

$$\mathbf{A} = \mathbf{A}^{\mathrm{T}}$$

and
$$A_{i,j} = A_{j,i} \quad \text{for all } i, j$$

An example of a symmetrical matrix would be the correlation matrix, commonly used as a starting point of many geostatistical methods. (Chapter 7)

Example 2.6
Show that the matrix C below, is symmetrical.

$$C = \begin{bmatrix} 1.000 & 0.569 & 0.766 & 0.326 & 0.010 & 0.284 \\ 0.569 & 1.000 & 0.528 & 0.267 & -0.04 & 0.123 \\ 0.766 & 0.528 & 1.000 & 0.688 & 0.169 & 0.678 \\ 0.326 & 0.267 & 0.688 & 1.000 & 0.210 & 0.669 \\ 0.010 & -0.04 & 0.169 & 0.210 & 1.000 & 0.381 \\ 0.284 & 0.123 & 0.678 & 0.669 & 0.381 & 1.000 \end{bmatrix}$$

Inspection of the matrix **C** shows that:

(a) it is symmetrical about the principal diagonal; and
(b) each row i is equal to the corresponding column j, where $i = j$

Thus if the matrix is rewritten as its transpose, it will be seen that the two are the same:

$$C^T = \begin{bmatrix} 1.000 & 0.569 & 0.766 & 0.326 & 0.010 & 0.284 \\ 0.569 & 1.000 & 0.528 & 0.267 & -0.04 & 0.123 \\ 0.766 & 0.528 & 1.000 & 0.688 & 0.169 & 0.678 \\ 0.326 & 0.267 & 0.688 & 1.000 & 0.210 & 0.669 \\ 0.010 & -0.04 & 0.169 & 0.210 & 1.000 & 0.381 \\ 0.284 & 0.123 & 0.678 & 0.669 & 0.381 & 1.000 \end{bmatrix}$$

The matrix **C** above is the correlation matrix for light hydrocarbon gases methane, ethene, ethane, propane, I–butane and N–butane, found in fluid inclusions from minerals in the Northern Pennine orefield, taken from unpublished data by the author.

Scalar operations defined for vectors also apply to matrices, thus if we multiply a matrix by a scalar, each element is multiplied by the scalar. Two matrices can be added (or subtracted) provided that they are of the same size. This follows since corresponding elements are added (or subtracted). Multiplication of a matrix by another matrix or by a vector is defined and will be dealt with in the next section, division on the other hand is not defined. Thus for example returning to the shorthand writing of a set of SLE:

$$A . x = z$$

is valid, though the algebraic solution:

$$x = \frac{z}{A}$$

is not valid since division is not defined in matrix algebra. The equivalent matrix operation for the solution of SLE is:

$$[A]^{-1} . z = x$$

where $[A]^{-1}$ is the **inverse** of the matrix **A**. Matrix inversion and the solution of SLE by this method will be dealt with later.

2.2.1 Matrix operations – multiplication by a vector

A matrix may be post multiplied by a vector provided that the number of **rows** of the vector is equal to the number of **columns** of the matrix. The result of the operation will be a vector with the same number of rows as the original matrix:

$$\mathbf{A}_{i,j} \cdot \mathbf{b}_k = \mathbf{x}_i$$

provided that $j = k$

where j is number of columns of the matrix

$\quad\quad k$ is the number of rows of the original vector

$\quad\quad i$ is the number of rows of the original matrix

$\quad\quad i$ is the number of rows of the resultant vector

This is the **row by column rule** for multiplication, and corresponds to the dot product of a pair of vectors. Thus the elements of row $a_{1,j}$ are multiplied by the corresponding elements b_k of the vector and summed to give element x_1 of the new vector until $a_{i,j}$, giving x_i. This is shown schematically in Fig. 2.1.

Example 2.7

Multiply together the following matrices and vectors, so that:

$$\mathbf{A} \cdot \mathbf{b} = \mathbf{c}$$

(a) $\quad \begin{bmatrix} 4 & 3 \\ 7 & 4 \end{bmatrix} \cdot \begin{bmatrix} 2 \\ 3 \end{bmatrix} = \begin{bmatrix} 4*2 + 3*3 \\ 7*2 + 4*3 \end{bmatrix} = \begin{bmatrix} 17 \\ 26 \end{bmatrix}$

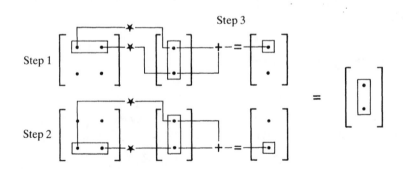

Figure 2.1 Schematic diagram representing the multiplication of a vector by a matrix.

(b) $\begin{bmatrix} 4 & 3 \\ 7 & 1 \end{bmatrix} \cdot \begin{bmatrix} 2 \\ 4 \\ 1 \end{bmatrix}$

This operation is not possible since j ≠ k

(c) $\begin{bmatrix} 4 & 3 \\ 7 & 1 \\ 2 & 4 \end{bmatrix} \cdot \begin{bmatrix} 2 \\ 3 \end{bmatrix} = \begin{bmatrix} 4 * 2 + 3 * 3 \\ 7 * 2 + 1 * 3 \\ 2 * 2 + 4 * 3 \end{bmatrix} = \begin{bmatrix} 17 \\ 17 \\ 16 \end{bmatrix}$

Example 2.8
Multiply together the following matrices and vectors so that:

$$\mathbf{b} \cdot \mathbf{A} = \mathbf{c}$$

(a) $[2 \ 3] \cdot \begin{bmatrix} 4 & 3 \\ 7 & 4 \end{bmatrix} = \begin{bmatrix} 2 * 4 + 3 * 7 \\ 2 * 3 + 3 * 4 \end{bmatrix} = \begin{bmatrix} 29 \\ 18 \end{bmatrix}$

(b) $[2 \ 3] \cdot \begin{bmatrix} 4 & 3 \\ 7 & 1 \\ 2 & 4 \end{bmatrix}$

This operation is not possible since j ≠ k

(c) $[2 \ 4 \ 1] \cdot \begin{bmatrix} 4 & 3 \\ 7 & 1 \\ 2 & 4 \end{bmatrix} = \begin{bmatrix} 2 * 4 + 4 * 7 + 1 * 2 \\ 2 * 3 + 4 * 1 + 1 * 4 \end{bmatrix} = \begin{bmatrix} 38 \\ 14 \end{bmatrix}$

Notice that the results for Example 2.7(a) and Example 2.8(a) differ, although the vector and matrix have the same numerical values. Also if **b** had been written as a column vector in Example 2.8(a), then the multiplication operation would not have been possible unless the matrix was post multiplied by the vector.

To sum up, for the operation:

$$\mathbf{A} \cdot \mathbf{b} = \mathbf{c}$$

where **A** is a square matrix with n rows and n columns
and **b** is a column vector with n rows
the resultant vector **c** will be a column vector with n rows.
The reverse multiplication:

$$\mathbf{b} \cdot \mathbf{A}$$

is impossible since the vector has only 1 column and the matrix n rows. However if we wrote **b** as a row vector with n columns and since the matrix has n rows, the multiplication would be possible.

Hence in matrix algebra pre- and post-multiplication are recognized, and as we shall show later, the result:

$$\mathbf{A} \cdot \# \neq \# \cdot \mathbf{A} \text{ (where \# is a vector or a matrix)}$$

holds in general (if the operation is possible). In most applications of matrix/vector multiplication, the matrix is post multiplied by the vector.

In an earlier example in this chapter the vector **w** which contained the number of ions in a sample of the mineral chlorite which had been analysed, was used without explanation as to its derivation. The starting point is the vector containing the analysis of the mineral in terms of the percentage of the oxides which after recalculating gives the vector **b** which contains the analysis in terms of moles of the oxides. The length (number of elements) of the vector will be j, where j is the number of oxides. We can also write a matrix **D**, the elements of which will be the number of atoms of each element i in oxide j. The matrix will have i rows and j columns (oxides) (where $i = j + 1$). Multiplication of the vector and the matrix gives the vector **w** which expresses the analysis in terms of the number of atoms in the original mineral, i.e.

$$\mathbf{D}(i, j) \cdot \mathbf{b}(j) = \mathbf{w}(i)$$

Example 2.9
Given the vector **b** containing the analysis of a chlorite in terms of the moles of oxide, find the vector **w** which contains the data in terms of atoms of the elements. The data is given in Table 2.2.

Table 2.2 Data for Example 2.9

Oxide	Percentage	Moles (b)
SiO_2	26.68	0.444
Al_2O_3	25.2	0.247
FeO	8.7	0.121
MgO	26.96	0.669
H_2O	11.7	0.649

Step 1   Draw up the matrix **D**, elements *vs* oxides:

$$
\begin{array}{c}
\\
\text{Si} \\
\text{Al} \\
\text{Fe}^{++} \\
\text{Mg} \\
\text{H}_2 \\
\text{O}
\end{array}
\begin{array}{c}
SiO_2 \quad Al_2O_3 \quad FeO \quad MgO \quad H_2O \\
\left[
\begin{array}{ccccc}
1 & 0 & 0 & 0 & 0 \\
0 & 2 & 0 & 0 & 0 \\
0 & 0 & 1 & 0 & 0 \\
0 & 0 & 0 & 1 & 0 \\
0 & 0 & 0 & 0 & 1 \\
2 & 3 & 1 & 1 & 1
\end{array}
\right]
\end{array}
$$

Step 2   Multiply the matrix **D** and the vector **b** to give the vector **w**:

		D				.	**b**		=	**w**	
	SiO_2	Al_2O_3	FeO	MgO	H_2O						
Si	1	0	0	0	0		0.444	SiO_2		0.444	Si
Al	0	2	0	0	0		0.247	Al_2O_3		0.494	Al
Fe^{++}	0	0	1	0	0	.	0.121	FeO =		0.121	Fe^{++}
Mg	0	0	0	1	0		0.669	MgO		0.669	Mg
H_2	0	0	0	0	1		0.649	H_2O		0.649	H_2
O	2	3	1	1	1					3.068	O

Note that the row and column labels have been left in to emphasize the relevance of position in vector and matrix layout.

It is important to notice that the matrix **D** has 6 rows and 5 columns and that the multiplier, vector **b** has 5 rows, and the resultant vector **w** has 6 rows.

2.2.2 Student examples

Q2.5 Perform the multiplication operations as indicated:

(a) $\begin{bmatrix} 2 & 3 & 6 \\ 1 & 5 & 2 \\ 6 & 2 & 1 \end{bmatrix} . \begin{bmatrix} 2 \\ 3 \\ 2 \end{bmatrix}$ (b) $\begin{bmatrix} -1 & 2 & 4 \\ 2 & 3 & 2 \\ 2 & 1 & -2 \end{bmatrix} . \begin{bmatrix} 10 \\ -1 \\ 4 \end{bmatrix}$

Q2.6 The vector of molecular proportions of oxides from an analysis of a hornblende is:

SiO_2	0.8593
Al_2O_3	0.0725
Fe_2O_3	0.0157
FeO	0.0738
MnO	0.0024
MgO	0.4487
CaO	0.2197
Na_2O	0.0098
H_2O	0.1283

Calculate the vector which expresses this information in terms of the atoms of each element in the mineral formulae on the basis of 24 oxygens.

Hint: The operations are:

$$\mathbf{D}_{(i,j)} . \mathbf{b}_{(j)} = \mathbf{w}_{(i)}$$
$$\mathbf{w}_{(i)} = \mathbf{w}_{(i)} * \frac{24}{w_{(ox)}}$$

The vector given is $\mathbf{b}_{(j)}$.

2.2.3 Multiplication of two matrices

Any pair of square matrices of the same order (size) can be multiplied together. The row by column rule defined for matrix/vector multiplication applies, the second matrix is treated as a number of column vectors, so that each resulting vector after multiplication, forms a column of the new matrix.

Example 2.10
Multiply together two matrices:

$$A . B = C$$

$$\begin{bmatrix} 4 & 3 \\ 7 & 4 \end{bmatrix} . \begin{bmatrix} 2 & 4 \\ 3 & 2 \end{bmatrix} = \begin{bmatrix} [(4 * 2) + (3 * 3)] & [(4 * 4) + (3 * 2)] \\ [(7 * 2) + (4 * 3)] & [(7 * 4) + (4 * 2)] \end{bmatrix}$$

$$C = \begin{bmatrix} 17 & 22 \\ 26 & 36 \end{bmatrix}$$

Note that the second matrix is another square matrix of the same order as the originals. The operation is shown schematically in Fig. 2.2.

Example 2.11
Perform the multiplication (**B** . **A**) using the same matrices as in Example 2.10:

$$B . A = D$$

$$\begin{bmatrix} 2 & 4 \\ 3 & 2 \end{bmatrix} . \begin{bmatrix} 4 & 3 \\ 7 & 4 \end{bmatrix} = \begin{bmatrix} [(2 * 4) + (4 * 7)] & [(2 * 3) + (4 * 4)] \\ [(3 * 4) + (2 * 7)] & [(3 * 3) + (2 * 4)] \end{bmatrix}$$

$$D = \begin{bmatrix} 36 & 22 \\ 26 & 17 \end{bmatrix}$$

which shows that in general:

$$(A . B) \neq (B . A)$$

Thus as noted earlier, multiplication in matrix algebra does not commute, thus post- and pre-multiplication are recognized. One exception to this rule is the operation involving the identity matrix, thus:

$$(A . I) = (I . A)$$

Matrices which have differing numbers of rows and columns may be multiplied together provided that the number of columns of the first matrix equals the number of rows of the second matrix, as in matrix/vector multiplication.

Thus if the first matrix has indices i, j and the second matrix has indices k, l then the first can be multiplied by the second provided that $j = k$ and the order of the resultant matrix will be i, l.

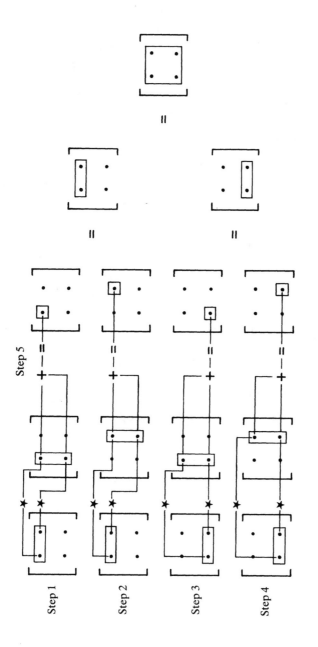

Figure 2.2 Schematic diagram representing matrix multiplication.

Example 2.12
Multiply the 2×3 matrix **A** by the 3×2 matrix **B**

$$\mathbf{A}_{2,3} \cdot \mathbf{B}_{3,2} = \mathbf{C}_{2,2}$$

$$\begin{bmatrix} 2 & 4 & 1 \\ 3 & 2 & 1 \end{bmatrix} \begin{bmatrix} 4 & 3 \\ 7 & 1 \\ 1 & 1 \end{bmatrix} = \begin{bmatrix} [(2*4)+(4*7)+(1*1)] & [(2*3)+(4*1)+(1*1)] \\ [(3*4)+(2*7)+(1*1)] & [(3*3)+(2*1)+(1*1)] \end{bmatrix}$$

$$\mathbf{C} = \begin{bmatrix} 37 & 11 \\ 27 & 12 \end{bmatrix}$$

The reverse multiplication is also possible since:

$$\text{for matrix } \mathbf{B} \qquad i = 3 \text{ and } j = 2$$
$$\text{and for matrix } \mathbf{A} \qquad l = 2 \text{ and } m = 3$$

the resulting matrix will be of order i, m (i.e. 3, 3)

$$\mathbf{B} \quad . \quad \mathbf{A} \quad = \quad \mathbf{D}$$

$$\begin{bmatrix} 4 & 3 \\ 7 & 1 \\ 1 & 1 \end{bmatrix} \cdot \begin{bmatrix} 2 & 4 & 1 \\ 3 & 2 & 1 \end{bmatrix} = \begin{bmatrix} 17 & 22 & 7 \\ 17 & 30 & 8 \\ 5 & 6 & 2 \end{bmatrix}$$

Example 2.13
Multiply the following matrices together:

$$\mathbf{A} = \begin{bmatrix} 2 & 4 & 1 \\ 3 & 2 & 1 \end{bmatrix} \qquad \mathbf{B} = \begin{bmatrix} 4 & 3 & 2 \\ 7 & 1 & 2 \end{bmatrix}$$

For **A**: $i = 2, j = 3$ for **B**: $l = 2, m = 3$

Therefore since $j \neq l$ the multiplication is not possible

Example 2.14
Multiply together the given matrices

$$\mathbf{A} = \begin{bmatrix} 2 & 4 & 1 \\ 3 & 2 & 1 \end{bmatrix} \qquad \mathbf{B} = \begin{bmatrix} 4 & 3 & 2 \\ 7 & 1 & 1 \\ 1 & 1 & 2 \end{bmatrix}$$

For **A**: $i = 2, j = 3$ for **B**: $l = 3, m = 3$

Therefore since $j = l$ the multiplication is possible and the resulting matrix will be of order i, m (i.e. 2, 3)

$$(\mathbf{A} \cdot \mathbf{B}) = \begin{bmatrix} 37 & 11 & 10 \\ 27 & 12 & 10 \end{bmatrix}$$

For the reverse multiplication, (**B** . **A**):
  Since the matrix **B** is 3, 3 and **A** is 2, 3
  and $j \neq l$ the multiplication is not possible.

In the discussion of the multiplication of two vectors only the dot product which resulted in a scalar was considered. However a vector is also a matrix, so that a row vector is a matrix with one row and n columns and a column vector is a matrix with n rows and one column. If the rule of matrix multiplication is applied then a column vector with n rows post multiplied by a row vector with n columns will result in an $n \times n$ matrix:

$$
\begin{bmatrix} y_1 \\ y_2 \\ \cdot \\ \cdot \\ y_n \end{bmatrix} \cdot \begin{bmatrix} x_1 & x_2 & \dots & x_n \end{bmatrix} = \begin{bmatrix} y_1x_1 & y_1x_2 & \cdot & \cdot & y_1x_n \\ y_2x_1 & y_2x_2 & \cdot & \cdot & y_2x_n \\ & & \cdot & & \\ & & \cdot & & \\ y_nx_1 & y_nx_2 & \cdot & \cdot & y_nx_n \end{bmatrix}
$$

Example 2.15
Multiply together the following vectors in the order given:

(a) $\begin{bmatrix} 2 \\ 7 \\ 1 \end{bmatrix} \cdot \begin{bmatrix} 3 & 1 & 4 \end{bmatrix} = \begin{bmatrix} 6 & 2 & 8 \\ 21 & 7 & 28 \\ 3 & 1 & 4 \end{bmatrix}$

(b) $\begin{bmatrix} 3 \\ 1 \\ -4 \end{bmatrix} \cdot \begin{bmatrix} 2 & 4 & 2 \end{bmatrix} = \begin{bmatrix} 6 & 12 & 6 \\ 2 & 4 & 2 \\ -8 & -16 & -8 \end{bmatrix}$

(c) $\begin{bmatrix} 2 & 4 & 2 \end{bmatrix} \cdot \begin{bmatrix} 3 \\ 1 \\ -4 \end{bmatrix} = (2 * 3) + (4 * 1) + (2 * -4) = 2$

Example 2.15 (c) emphasizes the fact that a row vector post multiplied by a column vector gives a scalar when the rules of matrix multiplication are applied.

2.2.4 Powering a matrix

A square matrix can be raised to any power n, by multiplying it by itself n-times (dot product). It should be clear from the earlier discussion relating to the part played by the row and column rule in matrix multiplication, matrices which are not square can not be powered.

Example 2.16
Find the third power of the matrix:

$$
\begin{bmatrix} 3 & 2 & 1 \\ 1 & 3 & 2 \\ 4 & 1 & 1 \end{bmatrix}
$$

$$\begin{array}{ccccc} [\mathbf{A}] & . & [\mathbf{A}] & = & [\mathbf{A}]^2 \end{array}$$

$$\begin{bmatrix} 3 & 2 & 1 \\ 1 & 3 & 2 \\ 4 & 1 & 1 \end{bmatrix} . \begin{bmatrix} 3 & 2 & 1 \\ 1 & 3 & 2 \\ 4 & 1 & 1 \end{bmatrix} = \begin{bmatrix} 15 & 13 & 8 \\ 14 & 13 & 9 \\ 17 & 12 & 7 \end{bmatrix}$$

$$\begin{array}{ccccc} [\mathbf{A}] & . & [\mathbf{A}]^2 & = & [\mathbf{A}]^3 \end{array}$$

$$\begin{bmatrix} 3 & 2 & 1 \\ 1 & 3 & 2 \\ 4 & 1 & 1 \end{bmatrix} . \begin{bmatrix} 15 & 13 & 8 \\ 14 & 13 & 9 \\ 17 & 12 & 7 \end{bmatrix} = \begin{bmatrix} 90 & 77 & 49 \\ 91 & 76 & 49 \\ 91 & 77 & 48 \end{bmatrix}$$

2.2.5 Student examples

Q2.7   Using the matrices given, show that:

$$(\mathbf{A} . \mathbf{B}) \neq (\mathbf{B} . \mathbf{A})$$

$$\mathbf{A} = \begin{bmatrix} 2 & 1 & 3 \\ 4 & 2 & 4 \\ 3 & 1 & 1 \end{bmatrix} \qquad \mathbf{B} = \begin{bmatrix} 4 & 0 & 1 \\ 2 & 3 & 2 \\ 2 & 2 & 1 \end{bmatrix}$$

Q2.8   Multiply the pairs of matrices in the order given. Is the reverse multiplication possible?

(a) $\mathbf{A} = \begin{bmatrix} 2 & 4 & 1 \\ 3 & 2 & 1 \end{bmatrix}$ $\mathbf{B} = \begin{bmatrix} 4 & 7 \\ 3 & 1 \\ 2 & 2 \end{bmatrix}$

(b) $\mathbf{A} = \begin{bmatrix} 2 & 4 & 1 \\ 3 & 2 & 1 \end{bmatrix}$ $\mathbf{B} = \begin{bmatrix} 4 & 3 & 2 \\ 7 & 1 & 1 \\ 1 & 1 & 2 \end{bmatrix}$

Q2.9   Multiply the following vectors in the order given.

(a) $\begin{bmatrix} 3 \\ 1 \end{bmatrix}$ $\begin{bmatrix} 2 & 4 \end{bmatrix}$

(b) $\begin{bmatrix} 2 \\ 1 \\ 3 \\ 1 \end{bmatrix}$ $\begin{bmatrix} -2 & 3 & 4 & 1 \end{bmatrix}$

(c) $\begin{bmatrix} 7 \\ 5 \\ 6 \end{bmatrix}$ $\begin{bmatrix} \alpha & \beta & 2 \end{bmatrix}$

Q2.10   Find the cube of the matrix:

$$\begin{bmatrix} 2 & 1 & 3 \\ 4 & 2 & 1 \\ 3 & 1 & 1 \end{bmatrix}$$

Does multiplying $[A]^2 . [A]$ rather than $[A] . [A]^2$, as in Example 2.16 make any difference to the result?

3

Determinants and inverses

In this chapter consideration of basic matrix operations is continued by looking at two very important topics in matrix algebra, those of the determinants of a matrix and the inverse of a matrix. Both of these lead to alternative techniques for the solution of SLE, but have wider implications as we shall see later. The value of both determinants and inverses in many advanced applications cannot be over stressed and students are advised to be sure that they are familiar with both topics before moving on.

3.1 Determinants

For every square matrix there is a specific scalar known as the determinant of the matrix. If the value of this scalar is zero then the matrix is said to be **singular**. Zero values are usually associated with matrices where linear dependency exists between the rows or the columns. The determinant is usually denoted by:

$$\det[A] \quad \text{or} \quad |A|$$

The determinant was first discovered during investigations into the solution of simultaneous linear equations and we shall look at that specific application of determinants later.

3.1.1 The calculation of determinants

If we have the general 2×2 matrix:

$$A = \begin{bmatrix} a & b \\ c & d \end{bmatrix}$$

then the determinant of A, is calculated as:

$$\det[A] = (a * d) - (c * b)$$

Example 3.1
Calculate determinants of the 2×2 matrices.

(a)
$$A = \begin{bmatrix} 4 & 3 \\ 7 & 1 \end{bmatrix}$$

then $\det[A] = (4 * 1) - (7 * 3) = -17$

(b)
$$B = \begin{bmatrix} 3 & 1 \\ 2 & 6 \end{bmatrix}$$

then $det[B] = (3 * 6) - (2 * 1) = 16$

(c)
$$C = \begin{bmatrix} 1 & 2.25 \\ 4 & 9 \end{bmatrix}$$

then $det[C] = 9 - 9 = 0$

(d)
$$D = \begin{bmatrix} 4 & 8 \\ 9 & 18 \end{bmatrix}$$

then $det[D] = 72 - 72 = 0$

The observant student will have noticed in Example 3.1 (c) that the rows of the matrix are multiples of one another and in Example 3.1 (d), that the columns are multiples of one another. This illustrates the rule that if any row (or column) of a matrix is a multiple of any other row (or column) there is a linear dependency and the determinant is zero. A matrix whose determinant is zero is known as a **singular matrix**. It should also be remembered that only square matrices have determinants.

If the matrix of coefficients of a set of linear equations has a determinant which is zero, then the equations as represented are not generally soluble, since they are not linearly independent.

We now require a rule which will allow us to calculate the determinant of a square matrix of any order. Consider the 3×3 matrix:

$$X = \begin{bmatrix} a_1 & b_1 & c_1 \\ a_2 & b_2 & c_2 \\ a_3 & b_3 & c_3 \end{bmatrix}$$

If one row and one column of the matrix are crossed off, for example row 1 and column 3, we are left with a 2×2 matrix:

$$\begin{array}{ccc} \cancel{a_1} & \cancel{b_1} & \cancel{c_1} \\ a_2 & b_2 & \cancel{c_2} \\ a_3 & b_3 & \cancel{c_3} \end{array}$$

which is:

$$\begin{bmatrix} a_2 & b_2 \\ a_3 & b_3 \end{bmatrix}$$

This matrix is a **minor** of the matrix $[X]$ and associated with this minor is the element $x_{1,3} = c_1$ (common to the row and column deleted).

Similarly if we cross off row 1 and column 2 we are left with the 2×2 minor:

$$\begin{bmatrix} a_2 & c_2 \\ a_3 & c_3 \end{bmatrix}$$

with its corresponding element $x_{1,2} = b_1$

finally, crossing off row 1 and column 1, we have the 2×2 minor:

$$\begin{bmatrix} b_2 & c_2 \\ b_3 & c_3 \end{bmatrix}$$

and its corresponding element $x_{1,1} = a_1$

Each minor can be signed, thus:

$$(-1) \quad \cdot \quad \begin{bmatrix} a_2 & c_2 \\ a_3 & c_3 \end{bmatrix}$$

the signed determinant is called a **cofactor** of the element $x_{1,2}$ ($= b_1$), and:

$$(+1) \quad \cdot \quad \begin{bmatrix} a_2 & b_2 \\ a_3 & b_3 \end{bmatrix}$$

the signed determinant is called a **cofactor** of the element $x_{1,3}$ ($= c_1$), and:

$$(+1) \quad \cdot \quad \begin{bmatrix} b_2 & c_2 \\ b_3 & c_3 \end{bmatrix}$$

the signed determinant is called a **cofactor** of the element $x_{1,1}$ ($= a_1$).

It is important to get the proper sign for the cofactor and the simplest way of achieving this is always to use row 1 for the elements, and to assign signs: $+ - + - \ldots - +$; starting at element 1, 1. Thus the sign of the cofactor is obtained from the sign of its associated element. The determinant of the matrix is calculated as the sum of the products of the cofactors and their elements, which for the 3×3 matrix above, is:

$$\det[\mathbf{X}] = a_1 \begin{vmatrix} b_2 & c_2 \\ b_3 & c_3 \end{vmatrix} - b_1 \begin{vmatrix} a_2 & c_2 \\ a_3 & c_3 \end{vmatrix} + c_1 \begin{vmatrix} a_2 & b_2 \\ a_3 & b_3 \end{vmatrix}$$

Note that for a 3×3 matrix the resulting cofactors are 2×2 minors whose determinants can be calculated as before. As we shall see presently, a 4×4 matrix will have four 3×3 minors, each of which has three 2×2 minors.

Example 3.2
Calculate the determinant of the 3×3 matrix **A**.

$$\mathbf{A} = \begin{bmatrix} 5 & 3 & 2 \\ 1 & 4 & 7 \\ 2 & 1 & 4 \end{bmatrix}$$

Step 1 Find the cofactors:

$$\det[A] = 5 \begin{vmatrix} 4 & 7 \\ 1 & 4 \end{vmatrix} - 3 \begin{vmatrix} 1 & 7 \\ 2 & 4 \end{vmatrix} + 2 \begin{vmatrix} 1 & 4 \\ 2 & 1 \end{vmatrix}$$

Step 2 Calculate the determinant of each minor, and multiply as appropriate:

$$= (5 * 9) - (3 * (-10)) + (2 * (-7))$$
$$= 45 + 30 - 14$$
$$= 61$$

Example 3.3

As a second example taking the matrix relating to grossular, kyanite and quartz to their component oxides as used earlier.

	gross	ky	qu
Al_2O_3	1	1	0
CaO	3	0	0
SiO_2	3	1	1

$$\det = 1 \begin{vmatrix} 0 & 0 \\ 1 & 1 \end{vmatrix} - 1 \begin{vmatrix} 3 & 0 \\ 3 & 1 \end{vmatrix} + 0 \begin{vmatrix} 3 & 0 \\ 3 & 1 \end{vmatrix}$$

$$= 1\,(0) - 1\,(3) + 0\,(3)$$

$$= -3$$

Example 3.4

Find the determinant of the matrix:

$$\begin{bmatrix} 5 & 14 & -14 \\ 14 & 40 & -34 \\ -14 & -34 & 73 \end{bmatrix}$$

$$\det = 5 \begin{vmatrix} 40 & -34 \\ -34 & 73 \end{vmatrix} - 14 \begin{vmatrix} 14 & -34 \\ -14 & 73 \end{vmatrix} + (-14) \begin{vmatrix} 14 & 40 \\ -14 & -34 \end{vmatrix}$$

$$= 5 * (1764) - 14 * (546) + (-14) * (84)$$

$$= 8820 - 7644 - 1176$$

$$= 0$$

The method of calculation outlined can be extended to any size of matrix, thus for a 4 × 4 matrix we will have to find the determinant of four 3 × 3 minors, each weighted by the elements of the top row (by analogy to the previous example 1, 1 will be positive, 1, 2 will be negative, 1, 3 will be positive and 1, 4 will be negative). Thus there is a rather long-winded calculation since each of the four 3 × 3 minors has three 2 × 2 minors.

i.e.

$$[\mathbf{A}] = \begin{bmatrix} a_1 & b_1 & c_1 & d_1 \\ a_2 & b_2 & c_2 & d_2 \\ a_3 & b_3 & c_3 & d_3 \\ a_4 & b_4 & c_4 & d_4 \end{bmatrix}$$

$$\det[\mathbf{A}] = a_1 \begin{vmatrix} b_2 & c_2 & d_2 \\ b_3 & c_3 & d_3 \\ b_4 & c_4 & d_4 \end{vmatrix} - b_1 \begin{vmatrix} a_2 & c_2 & d_2 \\ a_3 & c_3 & d_3 \\ a_4 & c_4 & d_4 \end{vmatrix} + c_1 \begin{vmatrix} a_2 & b_2 & d_2 \\ a_3 & b_3 & d_3 \\ a_4 & b_4 & d_4 \end{vmatrix} - d_1 \begin{vmatrix} a_2 & b_2 & c_2 \\ a_3 & b_3 & c_3 \\ a_4 & b_4 & c_4 \end{vmatrix}$$

$$= a_1 \left\{ b_2 \begin{vmatrix} c_3 & d_3 \\ c_4 & d_4 \end{vmatrix} - c_2 \begin{vmatrix} b_3 & d_3 \\ b_4 & d_4 \end{vmatrix} + d_2 \begin{vmatrix} b_3 & c_3 \\ b_4 & c_4 \end{vmatrix} \right\}$$

$$- b_1 \{ a_2 \dots \} + c_1 \{ a_2 \dots \}$$

$$- d_1 \{ a_2 \dots \}$$

Although the method just described is reliable in all situations it involves a large number of simple calculations. Most people can manage 4×4 matrices without much bother using a hand calculator, although when problems arise they are frequently associated with elements which have negative values. Although the average desk top computer can make light work of a 10×10 determinant using this method, the number of computations and hence time taken, increases rapidly as the size of the matrix grows. Fortunately there are a number of properties of determinants which can be utilized to speed up the calculations.

3.1.2 Cramer's rule

Before leaving the cofactor method of calculating the determinant it is convenient to consider Cramer's rule for the solution of simultaneous linear equations. Consider the general pair of equations:

$$ax_1 + bx_2 = z_1 \tag{1}$$

$$cx_1 + dx_2 = z_2 \tag{2}$$

Which can be written in terms of the unknowns x_1 and x_2 as:

$$x_1 = \frac{(z_1 * d) - (z_2 * b)}{(a * d) - (c * d)} \tag{3}$$

$$x_2 = \frac{(a * z_2) - (c * z_1)}{(a * d) - (c * d)} \tag{4}$$

It should be noted that the denominator of (3) and (4) is the determinant of the coefficient matrix. Thus providing that the determinant of this matrix is not zero and that z_1 and z_2 are not both equal to zero, then there will be a solution to the equations. The numerator of these two equations is also a determinant, the matrix being formed by replacing the vector of coefficients of the root being found, with the vector formed from the RHS of the equations.

This is Cramer's rule which states:

Provided the determinant of the coefficient matrix of a set of SLE is not zero, and the constants of the equations are not all zero, then a solution can be found as the ratio of two determinants.

The numerator is the determinant of the matrix formed by replacing the coefficients of the root being found by the constants from the RHS of the equations (for n unknowns there will be n matrices, one for each root), while the denominator is the determinant of the coefficient matrix itself.

Example 3.5

Use Cramer's rule to solve the following equations:

$$\begin{aligned} x + y - z &= 1 \\ 8x + 3y - 6z &= 1 \\ -4x - y + 3z &= 1 \end{aligned}$$

Step 1 Find the determinant of the coefficient matrix:

$$\begin{bmatrix} 1 & 1 & -1 \\ 8 & 3 & -6 \\ -4 & -1 & 3 \end{bmatrix} = \mathbf{A}$$

$$\det [\mathbf{A}] = 1 \begin{vmatrix} 3 & -6 \\ -1 & 3 \end{vmatrix} - 1 \begin{vmatrix} 8 & -6 \\ -4 & 3 \end{vmatrix} - 1 \begin{vmatrix} 8 & 3 \\ -4 & -1 \end{vmatrix}$$

$$= 3 - 0 - 4$$

$$= -1$$

As the determinant of the coefficient matrix is not zero, we can proceed.

Step 2 Form matrix $\mathbf{B}$ by replacing the coefficients of the x-terms in $\mathbf{A}$ by the constants, and find the determinant:

$$\begin{bmatrix} 1 & 1 & -1 \\ 1 & 3 & -6 \\ 1 & -1 & 3 \end{bmatrix} = \mathbf{B}$$

$$\det [\mathbf{B}] = 1 \begin{vmatrix} 3 & -6 \\ -1 & 3 \end{vmatrix} - 1 \begin{vmatrix} 1 & -6 \\ 1 & 3 \end{vmatrix} - 1 \begin{vmatrix} 8 & 3 \\ -4 & -1 \end{vmatrix}$$

$$= 3 - 9 + 4$$

$$= -2$$

Step 3   Form matrix **C** by replacing the coefficients of the *y*-terms in **A** by the constants, and find the determinant:

$$\begin{bmatrix} 1 & 1 & -1 \\ 8 & 1 & -6 \\ -4 & 1 & 3 \end{bmatrix} = \mathbf{C}$$

$$\det[\mathbf{C}] = 1 \begin{vmatrix} 1 & -6 \\ 1 & 3 \end{vmatrix} - 1 \begin{vmatrix} 8 & -6 \\ -4 & 3 \end{vmatrix} - 1 \begin{vmatrix} 8 & 1 \\ -4 & 1 \end{vmatrix}$$

$$= 9 - 0 - 12$$

$$= -3$$

Step 4   Form matrix **D** by replacing the coefficients of the *z*-terms in **A** by the constants, and find the determinant:

$$\begin{bmatrix} 1 & 1 & 1 \\ 8 & 3 & 1 \\ -4 & -1 & 1 \end{bmatrix} = \mathbf{D}$$

$$\det[\mathbf{D}] = 1 \begin{vmatrix} 3 & 1 \\ -1 & 1 \end{vmatrix} - 1 \begin{vmatrix} 8 & 1 \\ -4 & 1 \end{vmatrix} + 1 \begin{vmatrix} 8 & 3 \\ -4 & -1 \end{vmatrix}$$

$$= 4 - 12 + 4$$

$$= -4$$

Step 5   The roots can now be calculated as ratios of the determinants:

$$x = \frac{\det[\mathbf{B}]}{\det[\mathbf{A}]} = \frac{-2}{-1} = 2$$

$$y = \frac{\det[\mathbf{C}]}{\det[\mathbf{A}]} = \frac{-3}{-1} = 3$$

$$z = \frac{\det[\mathbf{D}]}{\det[\mathbf{A}]} = \frac{-4}{-1} = 4$$

Further examples of the use of Cramer's rule will be given at the end of this section in student examples.

3.1.3 Calculation of determinants by row reduction

The methods of calculation of determinants are now continued by listing a number of properties which can be utilized to speed up the process. Some of these properties can be verified simply by experimentation, the proof of others is beyond the scope of this text, and the student should consult one of the linear algebra texts listed in the references.

1. Writing a matrix as its transpose does not change the value of the determinant.

2. The interchange of a single pair of rows or columns will reverse the sign of the determinant.

3. If all the elements of a row or column are zero, then the determinant will be zero.

4. If any two rows or columns are identical, then the value of the determinant will be zero.

5. The multiplication or division of every element of a row or a column by a factor K is equivalent to multiplying or dividing the determinant by the same factor. It follows from this and 4 above, that if the corresponding elements of two rows or two columns are in the same proportion, then the determinant will be zero.

6. The addition or subtraction of corresponding elements of one row or column from those in any parallel row or column will not change the value of the determinant. It follows from this and from 4 above, that if the operation produced a new row or column which is identical to another row or column in the matrix, then the determinant will be zero.

7. The product of two determinants is equal to the determinant of the product of the two original matrices.

The application of these rules or properties should enable easier calculation of determinants by the method already described, after suitable manipulation of the matrix. They have also lead to a method of calculation based on row reduction. This is similar to the method described in Chapter 1 for the solution of SLE. The matrix is reduced to the **upper triangular form** and the determinant is found as the product of the elements in the principal diagonal The procedure is exactly the same as described in Chapter 1, section 1.2.1 as Gaussian elimination.

Example 3.6

Find the determinant of the coefficient matrix for a set of four reactions involving the minerals calcite, dolomite, diopside and forsterite (this is part of a thermodynamic problem which will be considered more fully at a later stage). The matrix is:

$$
\begin{array}{cccc}
\text{dol.} & \text{cal.} & \text{diop.} & \text{forst.} \\
\left[\begin{array}{cccc}
-11 & 13 & 0 & 1 \\
-2 & 2 & 0 & 1 \\
0 & -3 & 5 & 0 \\
0 & -1 & 2 & 0
\end{array}\right] &
\begin{array}{c}
(1) \\
(2) \\
(3) \\
(4)
\end{array}
\end{array}
$$

Step 1 Find a multiplier of row (1), which will eliminate element (1) from row (2). The multiplier is $\frac{-2}{11}$

	-11	13	0	1	(1)
multiply by				$\frac{-2}{11}$	

gives	2	$\frac{-26}{11}$	0	$\frac{-2}{11}$	
add	-2	2	0	1	(2)

gives	0	$\frac{-4}{11}$	0	$\frac{9}{11}$	(5)

The matrix is now:

$$\begin{bmatrix} -11 & 13 & 0 & 1 \\ 0 & \frac{-4}{11} & 0 & \frac{9}{11} \\ 0 & -3 & 5 & 0 \\ 0 & -1 & 2 & 0 \end{bmatrix} \begin{matrix} (1) \\ (5) \\ (3) \\ (4) \end{matrix}$$

Step 2 Find a multiplier for row (5) to eliminate element (2) from row (3). The multiplier is $\frac{-33}{4}$

	0	$\frac{-4}{11}$	0	$\frac{9}{11}$	(5)
multiply				$\frac{-33}{4}$	

gives	0	3	0	$\frac{-27}{4}$	
add	0	-3	5	0	(2)

gives	0	0	5	$\frac{-27}{4}$	(6)

Similarly eliminate element (2) from row (4):

	0	$\frac{-4}{11}$	0	$\frac{9}{11}$	(5)
multiply by				$\frac{-11}{4}$	

gives	0	1	0	$\frac{-9}{4}$	
add	0	-1	2	0	(2)

gives	0	0	2	$\frac{-9}{4}$	(7)

The matrix is now:

$$\begin{bmatrix} -11 & 13 & 0 & 1 \\ 0 & \frac{-4}{11} & 0 & \frac{9}{11} \\ 0 & 0 & 5 & \frac{-27}{4} \\ 0 & 0 & 2 & \frac{-9}{4} \end{bmatrix} \begin{matrix} (1) \\ (5) \\ (6) \\ (7) \end{matrix}$$

Step 4 Find a multiplier for row (6) to eliminate element (3) from row (7). The multiplier is $\frac{-2}{5}$

	0	0	5	$\frac{-27}{4}$	(6)
multiply by				$\frac{-2}{5}$	

| gives | 0 | 0 | -2 | $\frac{54}{20}$ | |
| add | 0 | 0 | 2 | $\frac{-9}{4}$ | (7) |

| gives | 0 | 0 | 0 | $\frac{9}{20}$ | (8) |

The matrix is now:

$$\begin{bmatrix} -11 & 13 & 0 & 1 \\ 0 & \frac{-4}{11} & 0 & \frac{9}{11} \\ 0 & 0 & 5 & \frac{-27}{4} \\ 0 & 0 & 0 & \frac{9}{20} \end{bmatrix}$$

(1)
(5)
(6)
(8)

Step 5 Calculate determinant by multiplying together elements of the principal diagonal:

$$\text{Det} = (-11) * \left(\tfrac{-4}{11}\right) * (5) * \left(\tfrac{9}{20}\right)$$
$$= 9$$

Calculating the determinant of the original matrix using the method of cofactors:

$$\det = -11 \begin{vmatrix} 2 & 0 & 1 \\ -3 & 5 & 0 \\ -1 & 2 & 0 \end{vmatrix} -13 \begin{vmatrix} -2 & 0 & 1 \\ 0 & 5 & 0 \\ 0 & 2 & 0 \end{vmatrix} -1 \begin{vmatrix} -2 & 2 & 0 \\ 0 & -3 & 5 \\ 0 & -1 & 2 \end{vmatrix}$$

$$\begin{vmatrix} 2 & 0 & 1 \\ -3 & 5 & 0 \\ -1 & 2 & 0 \end{vmatrix} = 2 \begin{vmatrix} 5 & 0 \\ 2 & 0 \end{vmatrix} -0 \begin{vmatrix} -3 & 0 \\ -1 & 0 \end{vmatrix} +1 \begin{vmatrix} -3 & 5 \\ -1 & 2 \end{vmatrix} = -1$$

$$\begin{vmatrix} -2 & 0 & 1 \\ 0 & 5 & 0 \\ 0 & 2 & 0 \end{vmatrix} = -2 \begin{vmatrix} 5 & 0 \\ 2 & 0 \end{vmatrix} -0 \begin{vmatrix} 0 & 0 \\ 0 & 0 \end{vmatrix} +1 \begin{vmatrix} 0 & 5 \\ 0 & 2 \end{vmatrix} = 0$$

$$\begin{vmatrix} -2 & 2 & 0 \\ 0 & -3 & 5 \\ 0 & -1 & 2 \end{vmatrix} = -2 \begin{vmatrix} -3 & 5 \\ -1 & 2 \end{vmatrix} -2 \begin{vmatrix} 0 & 5 \\ 0 & 2 \end{vmatrix} +0 \begin{vmatrix} 0 & -3 \\ 0 & -1 \end{vmatrix} = 2$$

$$\text{Det} = \left|(-11)(-1)\right| - \left|(13)(0)\right| - \left|(1)(2)\right|$$
$$= 9$$

Example 3.7

Find the determinant of:

$$\begin{bmatrix} 1 & 3 & 11 \\ -3 & 0 & -6 \\ 0 & 4 & 12 \end{bmatrix}$$

(1)
(2)
(3)

using the method of row reduction.
Exchange rows (2) and (3), to give:

$$\begin{bmatrix} 1 & 3 & 11 \\ 0 & 4 & 12 \\ -3 & 0 & -6 \end{bmatrix}$$

(1)
(2)
(3)

Eliminate element (1) from row (3):

	1	3	11	(1)
multiply by			3	

| gives | 3 | 9 | 33 | |
| add | -3 | 0 | -6 | (2) |

| gives | 0 | 9 | 27 | (4) |

The matrix is now:

$$\begin{bmatrix} 1 & 3 & 11 \\ 0 & 4 & 12 \\ 0 & 9 & 27 \end{bmatrix}$$

(1)
(3)
(4)

Eliminate element (2) from row (4):

	0	4	12	(3)
multiply by			$\frac{-9}{4}$	

| gives | 0 | -9 | -27 | |
| add | 0 | 9 | 27 | |

| gives | 0 | 0 | 0 | (5) |

The matrix becomes:

$$\begin{bmatrix} 1 & 3 & 11 \\ 0 & 4 & 12 \\ 0 & 0 & 0 \end{bmatrix}$$

(1)
(3)
(5)

From section 3.1.3 property 3, the determinant is zero, since all the elements of row (5) are zero.

3.1.4 Student examples

Q3.1 Calculate the determinants of the following matrices:

(a) $\begin{bmatrix} 2 & 3 \\ 2 & 4 \end{bmatrix}$ (b) $\begin{bmatrix} 3 & -1 \\ 1 & -1 \end{bmatrix}$ (c) $\begin{bmatrix} 1 & 5 & 3 \\ 2 & 4 & 1 \\ 3 & 3 & 6 \end{bmatrix}$

(d) $\begin{bmatrix} 3 & 2 & 1 \\ 1 & 3 & 1 \\ 3 & 9 & 3 \end{bmatrix}$ (e) $\begin{bmatrix} 1 & 3 & 2 \\ 2 & 1 & 0 \\ 3 & 6 & 1 \end{bmatrix}$

Q3.2 Using the matrices below, show that by changing over either two rows or two columns changes the sign of the determinant.

(b)

	gr	qu	woll
Al_2O_3	1	0	0
CaO	3	0	-1
SiO_2	3	1	-1

(c)

	gr	ky	qu
Al_2O_3	1	1	0
CaO	3	0	0
SiO_2	3	1	1

Using the same matrices, show that by changing another pair of rows or columns, changes the sign back. Try changing two columns for the set which you originally changed rows and vice versa.

Q3.3 Solve using Cramer's rule:

(a) $2x_1 + 3x_2 = 7$
     $2x_1 + 4x_2 = 8.5$

(b) $10A - 3B + C = 0$
     $16B - 3A + C = 0$
     $A + B - 100 = 0$

3.1.5 Determinants and linear independence

In section 1.3.4 the problem of linear independence was considered in the context of applying Gibbs' phase rule to chemical reactions in meta-morphic petrology. It was shown that if the solution of equation sets as homogeneous SLE was possible then there was no linear dependency among the reactions. Another way of viewing the same problem is by the use of determinants, since by the rules previously listed, the determinant of matrices where there is linear dependency will be zero (properties 3, 4, 5 and 6, section 3.1.3). This of course applies only to square matrices, and the problem noted in Chapter 1, that of underdetermined systems, still occurs. However there are properties of determinants and matrices which can be used to provide another solution to this problem.

From the properties of determinants we note that:

Property 1: $det[A] = det[A]^T$

Property 8: $\det([\mathbf{A}].[\mathbf{B}]) = \det[\mathbf{A}] * \det[\mathbf{B}]$

Also we can deduce from the general properties of matrices that the dot product of any matrix and its transpose will always be a square matrix (size dependent on the original, and the order of multiplication when the original is not square).

Therefore if an underdetermined equation set is considered, such as that quoted in Example 1.16, where we were required to ascertain whether or not the equilibrium relations were linearly independent, an alternative approach would be to find:

$$\det[\mathbf{A}]$$

where $$\mathbf{A} = \mathbf{C}.\mathbf{C}^{\mathrm{T}}$$

and $\mathbf{C}$ is the matrix of coefficients of the reactants and products. If linear dependence exists among the equations then the matrix $\mathbf{A}$ will be singular.

Example 3.8

Test the 3 × 4 matrix given, for linear independence:

$$\begin{bmatrix} 3 & 2 & 1 & 4 \\ 1 & 2 & 3 & 1 \\ 3 & 1 & 4 & 2 \end{bmatrix}$$

Step 1

$$\mathbf{A} \qquad . \qquad \mathbf{A}^{\mathrm{T}} \qquad = \qquad \mathbf{B}$$

$$\begin{bmatrix} 3 & 2 & 1 & 4 \\ 1 & 2 & 3 & 1 \\ 3 & 1 & 4 & 2 \end{bmatrix} \quad . \quad \begin{bmatrix} 3 & 1 & 3 \\ 2 & 2 & 1 \\ 1 & 3 & 4 \\ 4 & 1 & 2 \end{bmatrix} = \begin{bmatrix} 30 & 14 & 23 \\ 14 & 15 & 19 \\ 23 & 19 & 30 \end{bmatrix}$$

Step 2

$$\det[\mathbf{B}] = 30 \begin{vmatrix} 15 & 19 \\ 19 & 30 \end{vmatrix} - 14 \begin{vmatrix} 14 & 19 \\ 23 & 30 \end{vmatrix} + 23 \begin{vmatrix} 14 & 15 \\ 23 & 19 \end{vmatrix}$$

$$= 30(450 - 361) - 14(420 - 437) + 23(266 - 345)$$

$$= 2670 + 238 - 1817$$

$$= 1091$$

Since $\det[\mathbf{A}.\mathbf{A}^{\mathrm{T}}] = \det[\mathbf{B}] \neq 0$, then the matrix is linearly independent.

Example 3.9

Using the coefficients for the equilibrium relations given in Example 1.16, confirm the conclusion that there is linear dependency. The matrix of coefficients is:

$$C = \begin{bmatrix} 1 & 0 & 1 & 0 & -2 \\ 1 & 1 & 0 & -2 & 0 \\ 0 & 1 & -1 & -2 & +2 \end{bmatrix}$$

$$C.C^T = \begin{bmatrix} 6 & 1 & -5 \\ 1 & 6 & 5 \\ -5 & 5 & 10 \end{bmatrix}$$

$$\det[C.C^T] = 6 \begin{vmatrix} 6 & 5 \\ 5 & 10 \end{vmatrix} - 1 \begin{vmatrix} 1 & 5 \\ -5 & 10 \end{vmatrix} - 5 \begin{vmatrix} 1 & 6 \\ -5 & 5 \end{vmatrix}$$

$$= 0$$

and therefore the equilibrium relations are not linearly independent. Note: In the matrix $[C.C^T]$ if we reverse the sign of column 1 and add to column 2 we get column 3, which should tell us without the need for calculation, that the determinant is zero (property 6, section 3.1.3). This last example was taken from Spear, Rumble and Ferry (1982a, p.84), where the test for independence is given incorrectly, the multiplication operation being given in the incorrect order. The result given in the example they use (*ibid.*), is correct since the relationship:

$$\det[A^T.A] = 0$$

is always true, whether there is linear dependency or not.

3.1.6 Student examples

Q3.4 Using the matrix given in Example 3.8, show that the last statement in section 3.1.5 is true:

$$\text{i.e. } \det[C^T.C] = 0$$

Q3.5 Using the reaction considered in Example 1.13, show that there is no linear dependency. The matrix of coefficients is:

$$\begin{bmatrix} 1 & 1 & 0 & -1 \\ 3 & 0 & 0 & -1 \\ 3 & 1 & 1 & -2 \end{bmatrix}$$

3.2 Matrix inversion

If we consider any number x and its inverse $1/x$ then the product:

$$x * \frac{1}{x} = 1$$

similarly in matrix algebra the dot product of a square matrix and its inverse is the identity or unit matrix defined earlier. Note that **only square matrices have an inverse**. Thus:

$$[A] . [A]^{-1} = [I]$$

Example 3.10
Find the dot product of the matrix:

$$\begin{bmatrix} 4 & 2 \\ 6 & 5 \end{bmatrix} \text{ and its inverse, which is: } \begin{bmatrix} 0.625 & -0.25 \\ -0.75 & 0.5 \end{bmatrix}$$

$$\begin{bmatrix} 4 & 2 \\ 6 & 5 \end{bmatrix} \cdot \begin{bmatrix} 0.625 & -0.25 \\ -0.75 & 0.5 \end{bmatrix} = \begin{bmatrix} (2.5 - 1.5) & (-1 + 1) \\ (3.75 - 3.75) & (-1.5 + 2.5) \end{bmatrix}$$

$$= \begin{bmatrix} 1 & 0 \\ 0 & 1 \end{bmatrix}$$

The calculation of the inverse of a square matrix resolves itself into finding the matrix which when post multiplied by the matrix whose inverse is required, gives the identity matrix.

3.2.1 Matrix inversion by systematic elimination

In broad terms the method of systematic elimination involves the following operations:

1. Augment the matrix with an identity matrix of the same size, on the RHS of the original.
2. Using simple row operations to reduce the matrix to the identity matrix and, by performing the same operations on the original identity matrix, the required inverse will appear on the RHS.
3. Check calculations using the relationship:

$$(\mathbf{X} \cdot \mathbf{X}^{-1}) = (\mathbf{X}^{-1} \cdot \mathbf{X}) = \mathbf{I}$$

The process is best illustrated with a simple example.

Example 3.11
Find the inverse of the matrix:

$$\begin{bmatrix} 2 & 5 \\ 2 & 1 \end{bmatrix}$$

Step 1   Augment the matrix with the identity matrix of the same size

$$\begin{array}{cc|cc} 2 & 5 & 1 & 0 \\ 2 & 1 & 0 & 1 \end{array} \qquad \begin{array}{c} (1) \\ (2) \end{array}$$

Step 2   Divide row (1) through by 2, so that the first pivot is equal to unity

$$\begin{array}{cc|cc} 1 & \frac{5}{2} & \frac{1}{2} & 0 \\ 2 & 1 & 0 & 1 \end{array} \qquad \begin{array}{c} (1) \\ (2) \end{array}$$

Step 3 Find a multiplier of row (1) which will allow the elimination of element 2, 1, the multiplier is -2

	1	$\frac{5}{2}$	$\frac{1}{2}$	0
multiply by				-2

	-2	-5	-1	0
add	2	1	0	1

	0	-4	-1	1	(2)
this gives:					
	1	$\frac{5}{2}$	$\frac{1}{2}$	0	(1)
	0	-4	-1	1	(2)

Step 4 Divide row (2) through by -4 so that the second pivot is equal to unity

This gives:	1	$\frac{5}{2}$	$\frac{1}{2}$	0	(1)
	0	1	$\frac{1}{4}$	$\frac{-1}{4}$	(2)

Step 5 Find a multiplier of row (2) which will allow elimination of element 1, 2, the multiplier is $\frac{-5}{2}$

	0	1	$\frac{1}{4}$	$\frac{-1}{4}$
multiply by				$\frac{-5}{2}$

gives	0	$\frac{-5}{2}$	$\frac{-5}{8}$	$\frac{5}{8}$
add	1	$\frac{5}{2}$	$\frac{1}{2}$	0

	1	0	$\frac{-1}{8}$	$\frac{5}{8}$	(1)
This gives:					
	1	0	$\frac{-1}{8}$	$\frac{5}{8}$	(1)
	0	1	$\frac{1}{4}$	$\frac{-1}{4}$	(2)

and the matrix on the right is the inverse required.

Test:

$$\begin{bmatrix} 2 & 5 \\ 2 & 1 \end{bmatrix} \cdot \begin{bmatrix} \frac{-1}{8} & \frac{5}{8} \\ \frac{1}{4} & \frac{-1}{4} \end{bmatrix} = \begin{bmatrix} 1 & 0 \\ 0 & 1 \end{bmatrix}$$

There is a short-cut method which can be used to calculate the inverse of a 2×2 matrix. The procedure can be summarised:

Step 1 Find the determinant of the matrix.

Step 2 Interchange the NW and SE elements.

Step 3 Reverse the signs of the NE and SW elements.

Step 4 Divide each element by the determinant of the original matrix.

The student is left to verify that the inverses of the matrices used in Examples 3.10 and 3.11 are as quoted, using the short-cut method outlined.

It should be noted that the product of the divisors used to reduce the pivots to unity at each stage is equal to the determinant of the matrix (in the last example $2 * -4 = -8$, which is the determinant of the matrix). This is similar to the method of calculating the determinant by row reduction. Also it should be apparent from the short-cut method outlined for a 2×2 matrix, that if the matrix is singular it has no inverse.

Example 3.12
Find the inverse of the matrix:

$$\begin{bmatrix} 3 & 2 & 1 \\ 4 & 3 & 2 \\ 2 & 2 & 1 \end{bmatrix}$$ (1)
(2)
(3)

Augment the matrix and divide (1) through by 3, so that the first pivot is unity:

$$\begin{array}{ccc|ccc} 1 & \frac{2}{3} & \frac{1}{3} & \frac{1}{3} & 0 & 0 \\ 4 & 3 & 2 & 0 & 1 & 0 \\ 2 & 2 & 1 & 0 & 0 & 1 \end{array}$$ (1)
(2)
(3)

Multiply (1) by -4 to eliminate element 2, 1

$$\begin{array}{cccccc} 1 & \frac{2}{3} & \frac{1}{3} & \frac{1}{3} & 0 & 0 \end{array}$$

multiply by -4

$$\begin{array}{cccccc} -4 & \frac{-8}{3} & \frac{-4}{3} & \frac{-4}{3} & 0 & 0 \\ 4 & 3 & 2 & 0 & 1 & 0 \end{array}$$

add

$$\begin{array}{cccccc} 0 & \frac{1}{3} & \frac{2}{3} & \frac{-4}{3} & 1 & 0 \end{array}$$ (2)

Multiply (1) by -2 to eliminate element 3, 1

$$\begin{array}{cccccc} 1 & \frac{2}{3} & \frac{1}{3} & \frac{1}{3} & 0 & 0 \end{array}$$

multiply by -2

$$\begin{array}{cccccc} -2 \cdot & \frac{-4}{3} & \frac{-2}{3} & \frac{-2}{3} & 0 & 0 \\ 2 & 2 & 1 & 0 & 0 & 1 \end{array}$$

add

$$\begin{array}{cccccc} 0 & \frac{2}{3} & \frac{1}{3} & \frac{-2}{3} & 0 & 1 \end{array}$$ (3)

The augmented matrix is now:

$$
\begin{array}{ccc|ccc}
1 & \frac{2}{3} & \frac{1}{3} & \frac{1}{3} & 0 & 0 \\
0 & \frac{1}{3} & \frac{2}{3} & \frac{-4}{3} & 1 & 0 \\
0 & \frac{2}{3} & \frac{1}{3} & \frac{-2}{3} & 0 & 1
\end{array}
\qquad
\begin{array}{c}
(1) \\
(2) \\
(3)
\end{array}
$$

Divide (2) through by 1/3, so that the second pivot is unity:

$$
\begin{array}{cccccc}
0 & 1 & 2 & -4 & 3 & 0
\end{array}
\qquad (2)
$$

Multiply (2) by $-2/3$ to eliminate element 3, 2

$$
\begin{array}{cccccc}
0 & 1 & 2 & -4 & 3 & 0
\end{array}
$$

multiply by $\qquad\qquad\qquad\qquad\quad \frac{-2}{3}$

$$
\begin{array}{c}
 \\
\text{add}
\end{array}
\begin{array}{cccccc}
0 & \frac{-2}{3} & \frac{-4}{3} & \frac{8}{3} & -2 & 0 \\
0 & \frac{2}{3} & \frac{1}{3} & \frac{-2}{3} & 0 & 1
\end{array}
$$

$$
\begin{array}{cccccc}
0 & 0 & -1 & 2 & -2 & 1
\end{array}
\qquad (3)
$$

Divide row (3) by -1 so that the third pivot is unity

$$
\begin{array}{cccccc}
0 & 0 & 1 & -2 & 2 & -1
\end{array}
\qquad (3)
$$

Multiply row (3) by -2 to eliminate element 2, 3

$$
\begin{array}{cccccc}
0 & 0 & 1 & -2 & 2 & -1
\end{array}
$$

multiply by $\qquad\qquad\qquad\qquad\quad -2$

$$
\begin{array}{c}
 \\
\text{add}
\end{array}
\begin{array}{cccccc}
0 & 0 & -2 & 4 & -4 & 2 \\
0 & 1 & 2 & -4 & 3 & 0
\end{array}
$$

$$
\begin{array}{cccccc}
0 & 1 & 0 & 0 & -1 & 2
\end{array}
\qquad (2)
$$

Multiply (3) by $-1/3$ to eliminate element 1, 3

$$
\begin{array}{cccccc}
0 & 0 & 1 & -2 & 2 & -1
\end{array}
$$

multiply by $\qquad\qquad\qquad\qquad\quad \frac{-1}{3}$

$$
\begin{array}{c}
 \\
\text{add}
\end{array}
\begin{array}{cccccc}
0 & 0 & \frac{-1}{3} & \frac{2}{3} & \frac{-2}{3} & \frac{1}{3} \\
1 & \frac{2}{3} & \frac{1}{3} & \frac{1}{3} & 0 & 0
\end{array}
$$

$$
\begin{array}{cccccc}
1 & \frac{2}{3} & 0 & 1 & \frac{-2}{3} & \frac{1}{3}
\end{array}
\qquad (1)
$$

Multiply (2) by $-2/3$ to eliminate element 1, 2

	0	1	0	0	-1	2
multiply by						$\frac{-2}{3}$

	0	$\frac{-2}{3}$	0	0	$\frac{2}{3}$	$\frac{-4}{3}$
add	1	$\frac{2}{3}$	0	1	$\frac{-2}{3}$	$\frac{1}{3}$

	1	0	0	1	0	-1	(1)

Which gives, finally:

$$
\begin{array}{ccc|ccc}
1 & 0 & 0 & 1 & 0 & -1 \\
0 & 1 & 0 & 0 & -1 & 2 \\
0 & 0 & 1 & -2 & 2 & -1
\end{array}
\qquad
\begin{array}{c}
(1) \\ (2) \\ (3)
\end{array}
$$

and the required inverse is on the RHS.
Note that:

$$
\begin{bmatrix} 3 & 2 & 1 \\ 4 & 3 & 2 \\ 2 & 2 & 1 \end{bmatrix} \cdot \begin{bmatrix} 1 & 0 & -1 \\ 0 & -1 & 2 \\ -2 & 2 & -1 \end{bmatrix} = \begin{bmatrix} 1 & 0 & 0 \\ 0 & 1 & 0 \\ 0 & 0 & 1 \end{bmatrix}
$$

The determinant of the original matrix can now be calculated as the product of the divisors used to reduce the pivots to unity:

$$3 * \tfrac{1}{3} * -1 = -1$$

Check:

$$
\det = 3 \begin{vmatrix} 3 & 2 \\ 2 & 1 \end{vmatrix} - 2 \begin{vmatrix} 4 & 2 \\ 2 & 1 \end{vmatrix} + 1 \begin{vmatrix} 4 & 3 \\ 2 & 2 \end{vmatrix}
$$

$$= 3\,(-1) - 2\,(0) + 1\,(2)$$

$$= -1$$

One of the more important applications of matrix inversion is its use in the solution of SLE. Earlier (on page 37) we noted that a set of SLE could be solved in matrix form as:

$$\mathbf{A}^{-1} . \mathbf{c} = \mathbf{x}$$

In words: if the inverse of the coefficient matrix of a set of simultaneous linear equations, is post multiplied by the vector of constants, the result will be a vector containing the unknowns $x_1, x_2, \ldots, x_n$.

Example 3.13
The matrices used in Examples 3.11 and 3.12 are the coefficients of sets of SLE which are:

(a) $2x_1 + 5x_2 = 5.5$
   $2x_1 + 1x_2 = -0.5$

and (b) $3x_1 + 2x_2 + 1x_3 = 4$
$4x_1 + 3x_2 + 2x_3 = 7$
$2x_1 + 2x_2 + 1x_3 = 3$

Using the inverse matrices calculated earlier, find the roots for the two sets of equations.

(a)

$$\begin{bmatrix} \frac{-1}{8} & \frac{5}{8} \\ \frac{1}{4} & \frac{-1}{4} \end{bmatrix} \cdot \begin{bmatrix} 5.5 \\ -0.5 \end{bmatrix} = \begin{bmatrix} x_1 \\ x_2 \end{bmatrix}$$

$$= \begin{bmatrix} (-0.6875) + (-0.3125) \\ (1.375) + (0.125) \end{bmatrix}$$

$$= \begin{bmatrix} -1 \\ 1.5 \end{bmatrix}$$

Hence the roots are: $x_1 = -1$ and $x_2 = 1.5$

(b)

$$\begin{bmatrix} 1 & 0 & -1 \\ 0 & -1 & 2 \\ -2 & 2 & -1 \end{bmatrix} \cdot \begin{bmatrix} 4 \\ 7 \\ 3 \end{bmatrix} = \begin{bmatrix} x_1 \\ x_2 \\ x_3 \end{bmatrix}$$

$$= \begin{bmatrix} (4 - 3) \\ (-7 + 6) \\ (-8 + 14 - 3) \end{bmatrix}$$

$$= \begin{bmatrix} 1 \\ -1 \\ 3 \end{bmatrix}$$

Hence the roots are: $x_1 = 1$; $x_2 = -1$; $x_3 = 3$

3.2.2 Student examples

Q3.6 Find the determinants and the inverses of the following matrices:

(a) $\begin{bmatrix} 3 & 1 \\ 2 & 5 \end{bmatrix}$ (b) $\begin{bmatrix} 2 & 7 \\ 1 & 8 \end{bmatrix}$ (c) $\begin{bmatrix} 5 & 1 \\ 3 & 2 \end{bmatrix}$ (d) $\begin{bmatrix} 3 & 1 & 1 \\ 2 & 1 & 3 \\ 2 & 2 & 1 \end{bmatrix}$

Hint: Calculate (a)–(c) to 3 dp, leave (d) as fractions.

Q3.7 Find the inverse and determinant of:

$$\begin{bmatrix} 1 & 3 & 11 \\ -3 & 0 & -6 \\ 0 & 4 & 12 \end{bmatrix}$$

check the value of the determinant with the answer to Example 3.7.

Q3.8   Solve by matrix inversion:

$$\begin{aligned}
x + y - z &= 1 \\
8x + 3y - 6z &= 1 \\
-4x - y + 3z &= 1
\end{aligned}$$

Check your answer with that given for Example 3.5

Q3.9   Solve by matrix inversion:

$$\begin{aligned}
1x_1 + 2x_2 + 4x_3 &= -5523 \\
2x_1 + 1x_2 + 2x_3 &= -4038 \\
2x_1 \qquad\ + 3x_3 &= -4269.8
\end{aligned}$$

Check your answer with that given for Example 1.5

4

Accuracy and error

In the previous two chapters a number of important methods in linear algebra have been considered and their relationship to the problem of solving simultaneous linear equations has been outlined. The utility of the methods has been demonstrated and some possible application to a number of geological problems has been explored. An area which has not been discussed previously is the viability of the methods in practical situations, or the possible difficulties which might be encountered due to loss of accuracy because of round-off error during calculation. Also, although the two extreme situations where equations are multiples of one another or where they differ only by the constant have been noted, the consequences in the event that equations are almost multiples of one another, or that the constants differ by only a very small amount, have not been dealt with.

These problems are the subject of this chapter. It is not an exhaustive discussion, but hopefully it is sufficiently detailed to allow likely areas where problems can arise to be recognized. Some possible solutions will be suggested, which if followed can minimize their effect.

4.1 Examples of commonly occurring problems

To illustrate the difficulties which can arise during the solution of problems in linear algebra, two common examples will be considered. Both involve the process of matrix inversion, but could equally apply in other areas.

Example 4.1
Find the inverse of the given matrix, correct to 3dp

$$[A] = \begin{bmatrix} 5 & 4 & 2 \\ 6 & 4 & 3 \\ 0 & 2 & 5 \end{bmatrix}$$

The inverse was calculated, giving:

$$[A]^{-1} = \begin{bmatrix} -0.54 & 0.4 & -0.154 \\ 1.15 & -0.6 & 0.115 \\ -0.46 & 0.24 & 0.154 \end{bmatrix}$$

and the determinant as the product of the divisors:

$$\det[\mathbf{A}] = 5 * (-0.8) * 5 * 1.3 = -26$$

The calculation was checked:

$$[\mathbf{A}] \cdot [\mathbf{A}]^{-1} = [\mathbf{I}] = \begin{bmatrix} 0.980 & 0.080 & -0.002 \\ -0.020 & 0.720 & -0.002 \\ 0 & 0 & 1.00 \end{bmatrix}$$

and, the determinant:

$$\det[\mathbf{A}] = 5 \begin{vmatrix} 4 & 3 \\ 2 & 5 \end{vmatrix} -4 \begin{vmatrix} 6 & 3 \\ 0 & 5 \end{vmatrix} +2 \begin{vmatrix} 6 & 4 \\ 0 & 2 \end{vmatrix}$$

$$= -26$$

The differences between [**I**] and the product [**A**] . [**A**]$^{-1}$ were attributed to round-off error during calculation, as it was thought that insufficient significant figures had been used. However if the product [**A**]$^{-1}$. [**A**] had been calculated, one would not have been so complacent:

$$[\mathbf{A}]^{-1} \cdot [\mathbf{A}] = \begin{bmatrix} -0.300 & -0.868 & -0.650 \\ 2.150 & 2.430 & 1.075 \\ -0.860 & -0.572 & 0.570 \end{bmatrix}$$

and the result quoted would not have been accepted without further checking. The result, correct to 3 dp should have been:

$$[\mathbf{A}]^{-1} = \begin{bmatrix} -0.538 & 0.615 & -0.154 \\ 1.154 & -0.962 & -0.115 \\ -0.462 & 0.385 & 0.154 \end{bmatrix}$$

The problem had occurred during calculation, when the second pivot was driven to unity, for some reason, carelessness, the second element of the matrix which forms the inverse, was multiplied rather than divided. This of course does not effect the calculation of the determinant, but since row 2 is used to eliminate element 3, 2 and element 1, 2 all the subsequent values in row 2 of the inverse are affected.

As a second example, the problem of round-off error will be illustrated using fractions and decimals to 4 figure accuracy, inverting the coefficient matrix for a set of SLE and using the inverse to find the roots of the equation.

Example 4.2
Find the inverse of the matrix:

$$[\mathbf{A}] = \begin{bmatrix} 11 & 3 & 5 \\ 6 & 1 & 3 \\ 0 & 1 & 0 \end{bmatrix}$$

Using the method of systematic elimination, the inverses are:

$$[A]^{-1} = \begin{bmatrix} 1 & \frac{-5}{3} & \frac{-4}{3} \\ 0 & 0 & 1 \\ -2 & \frac{11}{3} & \frac{7}{3} \end{bmatrix} \text{ and } \begin{bmatrix} 1.012 & -1.686 & -1.349 \\ 0 & 0 & 1 \\ -2.024 & 3.705 & 2.364 \end{bmatrix}$$

The divisors are:

$$11, \frac{-7}{11}, \frac{3}{7} \text{ and } 11, -0.638, 0.423$$

and the determinants: -3 and -2.969

The original set of SLE, with roots $(5, -5, -2)$ were:

$$\begin{aligned} 11x + 3y + 5z &= 30 \\ 6x + y + 3z &= 19 \\ y &= -5 \end{aligned}$$

the roots x, y and z are given by:

$$A^{-1} . c = b$$

then:

$$\begin{bmatrix} 1.012 & -1.686 & -1.349 \\ 0 & 0 & 1 \\ -2.024 & 3.705 & 2.364 \end{bmatrix} . \begin{bmatrix} 30 \\ 19 \\ -5 \end{bmatrix} = \begin{bmatrix} 5.071 \\ -5 \\ -2.145 \end{bmatrix}$$

and for the inverse calculated using fractions:

$$\begin{bmatrix} 3 & -5 & -4 \\ 0 & 0 & 3 \\ -6 & 11 & 7 \end{bmatrix} . \begin{bmatrix} 30 \\ 19 \\ -5 \end{bmatrix} = \begin{bmatrix} 15 \\ -15 \\ -6 \end{bmatrix} * \frac{1}{3} = \begin{bmatrix} 5 \\ -5 \\ -2 \end{bmatrix}$$

Note: the inverse matrix was multiplied by 3 to eliminate the fractions and the resulting vector divided by 3 to give the final result.

The differences between the true values and those calculated from the decimal inverse, expressed as a percentage, are:

$$1.42\%$$
$$0\%$$
$$7.25\%$$

Although these differences may seem relatively small, they may become important if the values are used in further computations.

In the author's experience errors of this sort (round-off error) can be more problematic in practice, especially since most data processing is now performed by computer. An extreme example of this problem occurred during trend-surface computations. The program, run on an IBM 7094 computer (at that time – 1967, this was the 'state of the art' machine), was affected by round-off error during the computation of the sums of powers/sums of cross products matrix. The inverse of this matrix was then calculated, and then used in further calculations to obtain the coefficients of the equations on which the computed surfaces were to be based. In the event field measurements and mapping suggested that the structure being investigated was an anticline, the computer produced an interpretation which was the reverse, mapping a syncline. The problem lay in the large numerical values of the grid co-ordinates, taken from the National Grid without scaling. By arranging a new co-ordinate origin at the centre of the area being mapped, the problem was overcome. At a later date, the original data was processed using the same computer program on a machine with double the word length of the IBM 7094, this produced the correct interpretation, emphasizing the common problem relating the number of digits in a number being computed and the word length of the computer being used.

These examples illustrate the two principal sorts of error encountered during calculation. The first of these involves mistakes during hand calculation which can only be overcome with practice and care. Indeed the importance of careful layout of the problem at all stages, with clear crossing out when mistakes have been made, cannot be over-emphasized. These simple rules will go a long way toward preventing mistakes such as illustrated in Example 4.1. Errors in computer programs are beyond the scope of this book, but one word of advice might be appropriate – work through at least five examples by hand, with as wide a range of results as possible, and use these as test data to check the program. Experience suggests that although programme errors may yet remain undetected, this is still the best method – sadly human nature being what it is, many of us are satisfied if a single example produces the correct result. It is interesting to note that the iterative method of solution of equations outlined in Chapter 1, is self correcting and isolated arithmetic errors which may be made during hand calculation are corrected as the computation proceeds. The only obvious consequence being an increase in the number of iterations required to produce a result.

Round-off error is more pernicious, and must be guarded against particularly where coefficients are not simple integer values and when calculations are performed totally by hand or by the use of hand-held calculators. Typically, round-off error occurs with integer valued coefficients when the result of an operation is a number where there is no finite decimal representation. Examples are 1/3, 1/6, 1/11 etc. Such problems are compounded by differences between calculators in the way in which numbers are handled i.e. whether they are truncated or rounded. In certain extreme examples the difference can be critical. For example division by small numbers which themselves have been truncated or rounded.

Example 4.3

Divide 1.000 by the number 0.0000346, using it in its original form, then truncated and rounded to 6 d.p.

Operation	Result	Difference
$\dfrac{1}{0.0000346}$ =	28901.73	
$\dfrac{1}{0.000034}$ =	29411.76	+510.03
$\dfrac{1}{0.000035}$ =	28571.43	−330.30

4.2 Practical error detection during computation

The principal defence is being aware when such problems are likely to arise, so that necessary precautions can be taken to achieve the degree of accuracy required. This can also include changing the method of calculation used. The second line of defence is to monitor the accuracy as the calculation proceeds. We shall look at this aspect first, returning to the other later in this section.

4.2.1 The checksum method

Simple row operations as used for the solution of SLE and for matrix inversion, can easily be adapted to allow the monitoring of each stage of the calculation. The additional feature required for monitoring is the checksum. The checksum is formed by summing each element in a row according to its arithmetic sign, to give the **row total.** By treating these totals exactly the same as the other row elements, the results can be used to check the accuracy of the arithmetic. Thus for example if we divide a row through by the value of the pivot, we must also divide the checksum by the same value. Then, if a new checksum is calculated (for the new row elements), it should equal the old value divided by the pivot. The method will emphasize errors due to mistakes in calculation as well as loss of significant figures. Although the example used to demonstrate the technique involves matrix inversion, the reader should note that it can be applied in exactly the same way, to the solution of SLE by Gaussian elimination.

Example 4.4

Find the inverse of the coefficient matrix of the equations given. Check the accuracy of the calculation at each stage using checksums. Using the calculated inverse, find the roots of the equations.

$$3.159x_1 + 2.817x_2 + 1.259x_3 = 14.812$$
$$1.356x_1 + 2.347x_2 + 0.512x_3 = 7.439$$
$$0.534x_1 + 1.710x_2 + 2.656x_3 = 8.624$$

Step 1   The procedure adopted is exactly the same as outlined in Chapter 2 except that an extra column is placed to the right of the matrices, in which is noted the sum of all the elements in the row – the checksum.

						Σ	
3.159	2.817	1.259	1	0	0	8.235	(1)
1.356	2.347	0.512	0	1	0	5.215	(2)
0.534	1.710	2.656	0	0	1	5.900	(3)

The operations proceed as before, except that they **must** also be performed on the checksum.

Step 2   Divide row (1) through by 3.159, so that the first pivot is unity:

1	0.89174	0.39854	0.31656	0	0	2.606837	(1)

Step 3   Multiply (1) by −1.356 and add row (2), to eliminate element 2, 1

	1	0.89174	0.39854	0.31656	0	0	2.606837
multiply by							−1.356
	−1.356	−1.2092	−0.54042	−0.42926	0	0	−3.534871
add	1.356	2.347	0.512	0	1	0	5.215
	0	1.1378	−0.02842	−0.42926	1	0	1.680129

Step 3   Multiply (1) by −0.534 and add row (3), to eliminate element 3, 1

	1	0.89174	0.39854	0.31656	0	0	2.606837
multiply by							−0.534
	−0.534	−0.47619	−0.21282	−0.16904	0	0	−1.392051
add	0.534	1.710	2.656	0	0	1	5.900
	0	1.23381	2.44318	−0.16904	0	1	4.507949

Step 4   Having completed the above operations, **new** checksums can be obtained by summing the elements of each new row:

						Σ	
1	0.89174	0.39854	0.31656	0	0	2.60684	(1)
0	1.13780	−0.02842	−0.42926	1	0	1.68012	(2)
0	1.23381	2.44318	−0.16904	0	1	4.50795	(3)

By comparing the new values in the checksum column with those found during the row operations, any discrepancy between them should be obvious. Notice that the row sums calculated by the row operations are given to one more significant figure than that used in the calculation of the elements. The differences are:

$$\begin{bmatrix} 0.000003 \\ -0.000009 \\ 0.000001 \end{bmatrix} \quad (\Sigma\ 1)$$

Step 5 The next stage of the calculation is to drive the second pivot to unity and then eliminate element 3, 2.

Divide row (2) by 1.13780:

0	1	−0.02498		−0.37727	0.87889	0		1.476639	(2)

Multiply the resulting row (2), by −1.23381 and add to row (3), to eliminate the required element:

	0	1	−0.02498	−0.37727	0.87889	0	1.476639
multiply by							−1.23381
	0	−1.23381	0.03082	0.46548	−1.08438	0	−1.821892
add 0	0	1.23381	2.44318	−0.16904	0	1	4.50795
	0	0	2.47400	0.29644	−1.08438	1	2.686058

Step 6 Sum values for new rows to give new checksums:

							Σ	
1	0.89174	0.39854	0.31656	0	0		2.60684	(1)
0	1	−0.02498	−0.37727	0.87889	0		1.47664	(2)
0	0	2.47400	0.29644	−1.08438	1		2.68606	(3)

We can now find the differences between the checksums for rows (2) and (3) calculating as before, these are:

$$\begin{bmatrix} 0.000001 \\ 0.000002 \end{bmatrix} \quad (\Sigma\ 2)$$

Step 7 The calculation proceeds by driving the third pivot to unity, and then eliminating elements 2, 3 and 1, 3 (omitting details), and calculating checksums to give:

						Σ	
1	0.89174	0	0.26881	0.17468	0.16109	2.17414	(1)
0	1	0	−0.37428	0.86786	0.01010	1.50368	(2)
0	0	1	0.11982	−0.43831	0.40420	1.08571	(3)

With checksum differences:

$$\begin{bmatrix} -0.000005 \\ -0.000081 \\ 0.000000 \end{bmatrix} \quad (\Sigma\ 3)$$

Step 8 Finally, element 1, 2 is eliminated, to give:

$$
\begin{array}{ccc|cccc}
& & & & & & \Sigma \\
1 & 0 & 0 & 0.60257 & -0.59923 & -0.17010 & 0.83324 \quad (1)\\
0 & 1 & 0 & -0.37428 & 0.86786 & 0.01010 & 1.50368 \quad (2)\\
0 & 0 & 1 & 0.11982 & -0.43831 & 0.40420 & 1.08571 \quad (3)
\end{array}
$$

With a checksum difference for row 1, which is:

$$[-0.000008] \quad (\Sigma\ 4)$$

The required inverse correct to 3 dp is:

$$
\begin{bmatrix}
0.603 & -0.599 & -0.170 \\
-0.374 & 0.868 & 0.010 \\
0.120 & -0.438 & 0.404
\end{bmatrix}
$$

Step 9   Calculate the roots of the equations using:

$$[A]^{-1} \cdot c = x$$

where $\qquad c = \begin{bmatrix} 14.812 \\ 7.439 \\ 8.624 \end{bmatrix}$ whence $x = \begin{bmatrix} 3.01 \\ 1.004 \\ 2.003 \end{bmatrix}$

Students should bear in mind that the checksum method is useful as a guide to loss of significant figures as well as mistakes during the evaluation of individual elements. An inspection of the checksum differences for the example just worked (vectors labelled $(\Sigma\ 1) - (\Sigma\ 4)$) show that the greatest differences are in the order of 1 part in 15 000 (element 2 $(\Sigma\ 3)$) whilst the smallest is smaller than 1 part in 11 * 106 (element 3 $(\Sigma\ 3)$). Consideration of all of the checksum differences suggests that there may be problems associated with row 2 of the matrix, and care should be exercised if the values are to be used in further calculations.

4.2.2 The residual vector method

A second method, to be used to check the solution of SLE, is that whereby a **residual vector r** is found. This vector contains values which are the differences between the constants for the equations – vector c as given, and calculated constants c_{calc}, found by putting the values for the roots $x_1, \ldots,$ x_n, in the original equations. The operations are:

$$
\begin{aligned}
[A]^{-1} \cdot c &= x \\
A \cdot x &= c_{calc} \\
c - c_{calc} &= r
\end{aligned}
$$

Example 4.5

Using the results from Example 4.4, calculate the residual vector **r** and hence assess the accuracy of the calculations performed.

Step 1   The values for the roots of the equations have been calculated and given to 3 dp accuracy. These are put back into the original equations and the vector c_{calc} found:

$$
\begin{array}{ccc}
\mathbf{A} & \mathbf{x} & = & \mathbf{c_{calc}} \\
\begin{bmatrix} 3.159 & 2.817 & 1.259 \\ 1.356 & 2.347 & 0.512 \\ 0.534 & 1.710 & 2.656 \end{bmatrix} & \begin{bmatrix} 3.010 \\ 1.004 \\ 2.003 \end{bmatrix} = & \begin{bmatrix} 14.8586 \\ 7.4635 \\ 8.6441 \end{bmatrix}
\end{array}
$$

Step 2   The residual vector $\mathbf{r}$ can now be calculated:

$$
\begin{array}{cccc}
\mathbf{c} & - & \mathbf{c_{calc}} & = & \mathbf{r} \\
\begin{bmatrix} 14.812 \\ 7.439 \\ 8.624 \end{bmatrix} & - & \begin{bmatrix} 14.859 \\ 7.464 \\ 8.644 \end{bmatrix} & = & \begin{bmatrix} -0.047 \\ -0.025 \\ -0.020 \end{bmatrix}
\end{array}
$$

The residuals for this example show a decrease in their absolute values from r_1 to r_3 and would seem to indicate that there may have been loss of significant figures when forming the second pivot and using the values obtained in subsequent calculations. The result of applying both methods of checking the accuracy of the calculation to this particular set of equations, arrive at the same conclusion. That is: more significant figures should have been carried during operations leading to the formation of the second pivot.

There are three matters which should be emphasized at this point. First, although an example involving matrix inversion was used to demonstrate the checksum method, the technique can also be applied to the solution of SLE by Gaussian elimination in exactly the same manner (Example 4.6). Second, although in Example 4.2 the calculated roots were compared with the true value for the roots, this is not possible in practice, since we rarely know these values. The correct method is to calculate the residual vector for the constants. Finally, the checksum method is the more useful of the two, having wider application as well as picking up errors in the calculation as they occur, thus saving time and effort in the long run.

4.2.3 Scaling

In the earlier discussion the problem of error due to the use of either very large or very small numbers in a calculation, was mentioned. One way of negating these effects is by scaling. To this end there are a number of standard results which can be used to advantage should this problem arise, these are listed without verification. Among the student examples there is a question which allows a simple demonstration of the validity of these results.

1. If the **vector of constants** only is multiplied or divided by a factor, the roots of the equations will also be multiplied or divided by the same factor. The determinant is unchanged.
2. If the **vector of constants** and the **matrix of coefficients** are both multiplied or divided by a factor, the roots of the equations will be unchanged. The determinant will be multiplied or divided by the factor n-times, where n is the number or rows of the matrix. For example:
   if the true value of the determinant $= -1$
   the factor $= 5$
   and $n = 3$
   then the determinant of the scaled matrix will be:

$$-1 * 5 * 5 * 5 = -125$$

3. If the **matrix of coefficients only** is multiplied or divided by a factor, the roots of the equations will be divided or multiplied by the factor.
   i.e. The effect is reversed so that if we divide by a factor α, then the roots will be multiplied by the factor α.
   The determinant will be multiplied or divided by the factor n-times (not reversed), where n is the number or rows of the matrix, as in 2 above.
4. If a **single row (coefficients** and **constant)**, is multiplied or divided by a factor, the roots remain unchanged. The determinant will be multiplied or divided by the factor.
5. If the **nth column** or **the nth row** of the **coefficient matrix only,** is multiplied or divided by a factor, then the nth root only will be divided or multiplied by the factor – the effect is reversed as in 3 above. The determinant will be multiplied or divided by the factor (effect not reversed).

4.2.4 Pivoting

One method frequently recommended to circumvent division by small numbers which may have been either truncated or rounded, is to swop/rows or columns so that the largest absolute values are in the pivotal positions, this is known as pivoting. The equations do not need to be totally rearranged before solution commences, it is sufficient to make the largest coefficient the first pivot, and exchange further pivots if necessary as the calculation proceeds. The idea of row/column interchange is not new and several examples were demonstrated in Chapter 1, where rows were interchanged for ease of calculation. Students should remember:

A single exchange of a row or a column, changes the sign of the determinant.

Example 4.6
Solve the following set of equations using pivoting:

$$2x_2 + 4x_3 = 13.2$$
$$2x_1 + 1.5x_2 + x_3 = 21.64$$
$$x_1 + 6x_2 - 2x_3 = 26.62$$

Since the coefficient with the largest absolute value is the x_2-term in the third equation, row and column interchanges are performed so that the x_2-term is the first pivot (position 1, 1).

Step 1   Exchange columns 1 and 2

$$2x_2 + 4x_3 = 13.2$$
$$1.5x_2 + 2x_1 + x_3 = 21.64$$
$$6x_2 + 1x_1 - 2x_3 = 26.62$$

Step 2   Exchange rows 1 and 3. From now on a checksum column is also included in the calculation, principally to demonstrate its use during the solution of equations by Gaussian elimination.

		Σ	
$6x_2 + 1x_1 - 2x_3 = 26.62$	31.62		(1)
$1.5x_2 + 2x_1 + x_3 = 21.64$	26.14		(2)
$2x_2 + 4x_3 = 13.2$	19.20		(3)

Step 3   Eliminate x_2 from (2) and (3). In this example (1) will be divided instead of multiplying as has been usual in previous examples. This simple step will avoid the use of $-1/3$ when we come to eliminate the x_2-term from (3).

Divide (1) by $-6/1.5 = -4$:

			Σ
	$-1.5x_2 - 0.25x_1 + 0.5x_3 = -6.655$	-7.905
add	$1.5x_2 + 2x_1 + x_3 = 21.640$	26.14
	$1.75x_1 + 1.5x_3 = 14.985$	18.235

Divide (1) by $-6/2 = -3$

			Σ
	$-2x_2 - 0.3333x_1 + 0.6667x_3 = -8.8733$	-10.54
add	$2x_2 + 4x_3 = 13.2$	19.20
	$- 0.3333x_1 + 4.6667x_3 = 4.3267$	8.66

Which gives:

			Σ	
$6x_2 + 1x_1 - 2x_3 = 26.620$	31.62		(1)
$1.75x_1 + 1.5x_3 = 14.985$	18.235		(2)
$-0.3333x_1 + 4.6667x_3 = 4.3267$	8.6601		(3)

Notice that there is a small difference between the calculated checksum for (3) after eliminating x_2 and the new checksum obtained by adding the row elements of (3). The absolute difference is 0.0001.

The problem now is to eliminate one of the x-terms from (2) and (3). There are two possible ways to continue the solution of these equations – either use them as they are, or exchange rows and columns so that (3) is the pivotal equation and x_3 is the pivot. Before continuing let us examine the possibilities and their outcomes more closely. If we use the equations as they are then the divisor is:

$$\frac{1.75}{0.3333} = 5.2505$$

and by changing the pivots the divisor will be:

$$\frac{-4.6667}{1.5} = -3.1111$$

Clearly both of these calculations involve numbers which are inexact, in the first it is the denominator while in the second it is the numerator. In practice using the first option, where the denominator is rounded will introduce greater error than the other, therefore common sense would suggest that the largest pivot be used. Since these equations have exact roots, a direct comparison is possible and we will use both alternatives to demonstrate what happens. The relevant two equations in their original form are:

$$
\begin{array}{lll}
 & \Sigma & \\
1.75x_1 + \quad 1.5x_3 = 14.985 & 18.235 & (2) \\
-0.3333x_1 + 4.6667x_3 = \quad 4.3267 & 8.6601 & (3)
\end{array}
$$

Step 4 Eliminate x_1 from (3), by dividing (2) by: $1.75/0.3333 = 5.2505$

$$
\begin{array}{ll}
 & \Sigma \\
0.3333x_1 + 0.2857x_3 = 2.854 & 3.4730 \\
\text{add} \quad -0.3333x_1 + 4.6667x_3 = 4.3267 & 8.6601 \\
\hline
4.9524x_3 = 7.1807 & 12.1331
\end{array}
$$

The checksum obtained by adding the coefficient of the x_3-term and the constant for the last equation is:

$$4.9524 + 7.1807 = 12.1331$$

and the absolute difference between that and the calculated value is:

$$12.1331 - 12.1331 = 0.0$$

The roots of the equation can now be calculated to give:

$$
\begin{aligned}
x_3 &= 1.4494 \\
x_2 &= 3.7012 \\
x_1 &= 7.3205
\end{aligned}
$$

The equations (2) and (3) above can be solved again, but on this occasion with the largest absolute value in the pivotal position. Rewriting the equations we have:

$$
\begin{array}{rlll}
 & & & \Sigma \\
4.6667x_3 - 0.3333x_1 = & 4.3267 & 8.6601 & (2) \\
1.5 \quad x_3 + 1.75x_1 = & 14.985 & 18.235 & (3)
\end{array}
$$

Step 5   Eliminate x_3 from (3)
Divide (2) by $-4.6667/1.5 = -3.1111$

$$
\begin{array}{rll}
 & & \Sigma \\
-1.5x_3 + 0.1071x_1 = -1.3907 & -2.78361 \\
1.5x_3 + 1.75x_1 \quad = 14.985 & 18.235 \\
\hline
1.8571x_1 = \quad 13.5943 & 15.45139
\end{array}
$$

add

The checksum obtained by adding the coefficient of the x_1-term and the constant is 15.4514 which is an absolute difference of 0.00001, which is considerably smaller than the difference obtained earlier. From the last equation, using back substitution, the roots are:

$$
\begin{aligned}
x_1 &= 7.3202 \\
x_3 &= 1.4498 \\
x_2 &= 3.6999
\end{aligned}
$$

Residual vectors can be calculated as in Example 4.5, which written as row vectors for convenience, are:

$$
\begin{aligned}
(\mathbf{r})_1 &= (-0.0089 \quad -0.0022 \quad 0) \\
(\mathbf{r})_2 &= (0 \quad -0.00005 \quad 0.001)
\end{aligned}
$$

$(\mathbf{r})_1$ refers to the equations unchanged while $(\mathbf{r})_2$ refers to the calculation performed after changing the pivot. Clearly the effect of changing the pivot is quite marked and in fact has changed the accuracy of the calculation considerably. The true values for the roots of these equations are:

$$
\begin{aligned}
x_1 &= 7.32 \\
x_2 &= 3.70 \\
x_3 &= 1.45
\end{aligned}
$$

In this section the methods discussed enable problems of accuracy associated with the solution of well-behaved equation sets, to be tackled. Clearly there can be no single recommended method, but scaling excepted, the use of pivoting with the checksum column for both matrix inversion and the solution of SLE when the coefficients are inexact, has much to commend it particularly in so far as awkward arithmetic can be avoided and mistakes in calculation can be picked up as they occur, thus saving time and effort. Scaling is a matter of choice. Clearly if numbers in one particular row or column are large or small compared to the others, there is merit in adjusting them. Also, and in particular in the case of regression analysis, scaling of all the values (coefficients and constants) may be advisable. In the next section we will examine the characteristics of ill-conditioned equations, and consider what can be done to solve them.

4.2.5 Student examples

Q4.1 Solve using Gaussian elimination with checksums, the equation set:

$$25A + 12.3B + 12.7C = 43.876$$
$$12.3A + 7.93B + 6.25C = 21.164$$
$$12.7A + 6.25B + 8.57C = 22.017$$

Compare your answer with the solution obtained by using Cramer's rule to solve the same equations.

Q4.2 Solve the equations below as set out and using pivoting:

$$2.01x_1 + 0.8x_2 + 2.35x_3 = 8.489$$
$$10.07x_1 + 3.98x_2 + 11.80x_3 = 42.5323$$
$$1.50x_1 + 1.2x_2 + 4.10x_3 = 10.45$$

Compare your answers. The true roots are: (2.05, 1.26, 1.43).

Q4.3 Solve the following equations by Gaussian elimination:

$$8x + 3y - 6z = 1$$
$$x + y - z = 1$$
$$-4x - y + 3z = 1$$

Use the equations to demonstrate the truth of the scaling rules, given earlier. What effect do the operations have on the value of the determinant of the coefficient matrix?

(i)   Multiply the constants by 10.0
(ii) Divide the y-coefficients by 10.0
(iii) Multiply all the coefficients and constants by 5.0
(iv) Multiply the coefficients and constant of the third equation by 3.0
(v)   Divide all the elements of the coefficient matrix by 2.0

4.3 Ill-conditioned equations

In Chapter 1 it was noted that if a set of SLE differed only by the value of the constants or if equations were exact multiples of each other, there would be no solution. Plotted graphically, the first of these give rise to parallel lines whilst the second, plotting one on top of the other gives rise to a single line. In the situation where equations are almost exact multiples of one another, they will plot as sub-parallel lines which have a point of intersection at some extreme value, hence the accuracy to which the calculation is performed can have a considerable bearing on the results. Also, small variations in the constants can give rise to large variations in the roots. These points can best be illustrated with reference to an example.

Example 4.7
Solve the following equations using Cramer's rule:

$$1.436x_1 + 1.122x_2 = 1.237 \tag{1}$$
$$2.708x_1 + 2.120x_2 = 2.334 \tag{2}$$

Compare the result obtained with the values for the roots if the constants are:

(a) (1.227, 2.334) (b) (1.237, 2.336)

(i) Solution of the original pair of equations using Cramer's rule, gives:

$$x_1 = \frac{(1.237 * 2.120 - 2.334 * 1.122)}{(1.436 * 2.120 - 2.708 * 1.122)}$$

$$= \frac{0.003692}{0.005944} = 0.6211 \text{ (to 4 dp)}$$

$$x_2 = \frac{(1.436 * 2.334 - 2.708 * 1.237)}{0.005944}$$

$$= \frac{0.001828}{0.005944} = 0.3075 \text{ (to 4 dp)}$$

(ii) If the constants are: (1.227, 2.334)

Then:
$$x_1 = -2.9455$$
$$x_2 = 4.8634$$

If the constants are: (1.237, 2.336)

Then:
$$x_1 = 0.2436$$
$$x_2 = 0.7907$$

The results for this example are summarized in Table 4.1, and it should be noted that a change of the constant in (1) by 1 part in 100 (1.237 changed to 1.227) has produced a change in the values for the roots of about 16 to 1. While a change of the constant in (2) by 2 parts in 1000 (2.334 changed to 2.336) has resulted in a change in the roots of about 2.5 to 1.

It is also instructive while considering this example to compare the differences in the results obtained, using a number of different methods of calculation to solve the equations as originally specified. These results are summarized in Table 4.2.

Table 4.1 Roots of the equations used in Example 4.7, illustrating the effect of changing the values of the constants of the equations, by relatively small amounts

Constants	Equation (1)	1.237	1.227	1.237
	Equation (2)	2.334	2.334	2.336
Roots	x_1	0.6211	−2.9455	0.2436
	x_2	0.3075	4.8634	0.7907

Table 4.2 Calculated values for the roots of the equations in Example 4.7 with constants as originally specified, obtained using different methods of calculation (rounded to 4 dp)

Method	(1)	(2)	(3)	(4)	(5)	(6)
Root x_1	0.6211	0.6211	0.6258	0.6179	0.6209	0.6215
Root x_2	0.3076	0.3075	0.2718	0.3076	0.3076	0.3071

(1) Computer solution using matrix inversion. (2) Cramer's rule as demonstrated (6 dp). (3) Matrix inversion calculation to 4 dp accuracy. (4) Matrix inversion to 6 dp accuracy. (5) Quick matrix inversion for a 2×2 matrix to 6 dp accuracy. (6) Gauss–Seidel iteration using a computer.

This exercise demonstrates the variation in results which can arise from different approaches to the calculation. In this example the variations in the first root x_1 are most noticeable. Equations which exhibit this property, or that demonstrated earlier where small changes in the constants led to large changes in the roots evaluated, are termed ill-conditioned. In general, if ill-conditioned equations are suspected, then Cramer's rule is probably the best method to use for their solution (provided that the number of unknowns is not too large). The reason for this is simple, there is only one division operation per root, and since this is the final operation of the calculation, any inaccuracies from truncation or rounding are not carried forward. The tabulated result (6), obtained by Gauss–Seidel iteration using the method of successive corrections was produced after 3312 iterations with the tolerance set at 10^{-5} (other values of the tolerance gave answers which differed widely from those obtained by the other methods).

Having examined some of the difficulties which can be encountered, it is useful to summarize some of the more important characteristics of ill-conditioned equations, so that they may be recognized and appropriate action taken.

1. Small changes in either the constants or coefficients give rise to large variations in the values of the roots.

2. Different methods of solution can, on occasion, produce varying values for one or more of the roots.
3. If the Gauss–Seidel iteration is used, convergence is slow.
4. The absolute value of the determinant is small numerically, relative to the coefficients. In the last example the determinant is 0.0059.
5. The elements of the inverse of the matrix are large relative to the coefficients (students should satisfy themselves that this is true by inverting the coefficient matrix for Example 4.6).
6. The ratio of the largest to smallest eigenvalue (see later, Chapter 6), is large.

If ill-conditioned equations are suspected, or if in doubt and provided the number of roots is not too large, use Cramer's rule. When this is not practical then it is advisable to use the checksum method, carrying as large a number of significant figures as possible to ensure an accurate solution. Students should also note that for ill-conditioned equations the residuals if calculated, will not necessarily be large and they should not be used as evidence for judgements about the condition of the original equations.

4.3.1 Student example

Q4.4 Solve the following equations by Gaussian elimination:

$$2.01x + 0.8y + 2.35z = 8.5$$
$$10.07x + 3.98y + 11.8z = 42.5323$$
$$1.50x + 1.20y + 4.1z = 10.45$$

Compare your result with that obtained for Q4.2 and from 4 above, that if the operation produced a new row or column which is identical to another row or column.

5

A geometrical interpretation of vector and matrix operations

In Chapters 1 and 2 a largely algebraic approach to linear algebra has been used to illustrate the basic operations. In particular emphasis has been placed on the properties and the solution of simultaneous linear equations. Indeed the reader might be forgiven for assuming that linear algebra is solely concerned with this particular problem. In this chapter the geometrical implications of the operations already discussed will be considered. One new topic, that of transformations will be introduced, essentially in a geometric setting.

Much of what will be dealt with concerns either angular relationships or transformations and in most instances 2–dimensional examples will be used by way of illustration. This applies particularly to simple deformation. The reason that 2–dimensional examples have been chosen is two-fold, simplifying the calculations and making illustration easier to draw and to interpret. The results demonstrated are equally valid in three or more dimensions, and this should be borne in mind when reading the text and working through examples. In the case of transformations a unit square will be used, to assist in keeping track of reflections and/or rotations, by reference to labelled corners.

5.1 Points and lines in the plane and in space

5.1.1 Vectors

Early in Chapter 2 scalars and vectors were defined in the context of the purpose for which they were needed. We shall now expand these definitions in order to be more specific. For example geologists often talk about the angle of dip of a bedding plane, simply by quoting the measured angle. This quantity is relatively meaningless unless accompanied by further information relating to the direction of the dip, or to the strike of the beds to which it refers. Thus dip is a **vector** quantity rather than a **scalar** quantity, since it has both magnitude and direction. Other examples of vector quantities which commonly occur in geological studies are displacement (of a fault for example), force and heat flow. The ratio of simple pairs of measurements can also have magnitude and direction, for example if the length and width of a brachiopod is measured and expressed as a ratio, this is a vector and should be treated as such.

In section 1.1 dealing with the solution of a pair of SLE the two roots of the equations were shown to be an ordered pair of numbers, which could be represented as a point with reference to a pair of mutually orthogonal axes. It was shown in Fig. 1.3 that this point was where the lines representing the two equations intersected. Similarly the roots of three equations in three unknowns can be represented as an ordered set of three numbers which represent a point in space with reference to three mutually orthogonal axes. Later, in section 2.1.2 a vector was defined as an ordered list of numbers in either row or column form, the position of a particular number having meaning in terms of the problem being tackled. Thus the roots of a set of SLE, is a particular example of a vector as defined in section 2.1.2. Further, since a vector can be represented by a line segment with reference to two or more mutually orthogonal axes, we can also represent it by a line drawn from the origin of the axes to the point specified. Such a line will have magnitude (in terms of its length) and direction (relative to the axes).

Among the examples used to demonstrate the methods of solution of SLE in Chapter 1 were problems in geochemistry where there were three or more variables (either oxides or minerals). In these examples and in some methods used in multivariate statistics which will be introduced in Chapter 7, each variable can be represented by an axis which is orthogonal with any other axis representing another variable. This, although it cannot be represented in a diagram, is an important concept.

There can be as many axes as required to represent the variables, each orthogonal with every other axis in the set.

Also there exists a row or column vector with as many elements as there are variables, which can be represented by a line drawn from the origin of the axes to the point which has coordinates identical to these elements. This is called the **position vector** of the point. This line will have both magnitude and direction with reference to the space defined by the particular problem under consideration.

Before looking in more detail at the mathematical treatment of vectors, we consider a number of specific examples which commonly arise in geology. Take for instance the row vector (5330 1801) which is the Grid reference of London Bridge railway station (to the nearest 1/10th km), and the row vector (390 540 450) which represents the thickness of a lithological unit measured at a specific point. The first of these is a vector relative to an origin in terms of two axes, the easting (first four digit number) and the northing (second four digit number). In the second example the first two numbers (three digits each) represent the vector (easting and northing as before), while the third represents the thickness (450 units) of the measured stratum, so that this is simply an example of a vector in two dimensions with additional information measured at a specific point. If on the other hand, in the second example the third number represented the depth measured from a datum to some stratum or event of geological interest, then this would be an example of a three-dimensional vector as defined earlier. As a further example we consider the row vector (1 0 1) which is the Miller index of a crystal face which cuts the x_1 axis at 1 unit, cuts the x_2 axis at 3 units (i.e. is parallel to x_2) and cuts the vertical x_3 axis at 3/2 units. Thus the Miller index (h k l) is the reciprocal of the intercepts a crystal face makes with the crystallographic axes and refers to a plane (the crystal face). As we have already indicated (section 1.3.4), the zone axis symbol $[U, V, W]$, represents a point rather than the plane as in the case of the Miller index. It should be noted that in the case of the Miller index, some properties allow it to be used as a vector.

In many physical systems it is convenient and advantageous to think of vectors in terms of their **magnitude** and **direction** with reference to their origin, conventionally taken as the point (0,0) or (0,0,0). In the case of the Grid reference quoted earlier, the origin is taken as a point to the south-west of the Isles of Scilly, while the zone axis and the related Miller index system, use as their reference the intersection of three imaginary crystallographic axes whose origin is at the centre of the crystal. As a final reminder, the roots $(x_1, x_2, \ldots, x_n)$ of a set of SLE, are understood to be evaluated with reference to the origin of the axes.

Using these concepts we can calculate the magnitude or length of the position vector of the point (x_1, x_2), using Pythagoras' theorem:

$$\text{Magnitude} = \sqrt{(x_1{}^2 + x_2{}^2)}$$

and the direction is given by:

$$\text{Direction} = \tan^{-1}\frac{x_1}{x_2}$$

as illustrated in Fig. 5.1a.

Example 5.1
Using the Grid reference quoted earlier for London Bridge station, find the distance and direction from the origin of the grid.

$$\text{Distance} = \sqrt{(533.0^2 + 180.1^2)} \text{ km}$$
$$= 526.606 \text{ km}$$
$$\text{Direction} = \tan^{-1}\frac{533}{180.1}$$
$$= \tan^{-1} 2.960$$
$$\fallingdotseq 71° \ 20'$$

The magnitude (length) of the position vector of the point (x_1, x_2, x_3) will be given as:

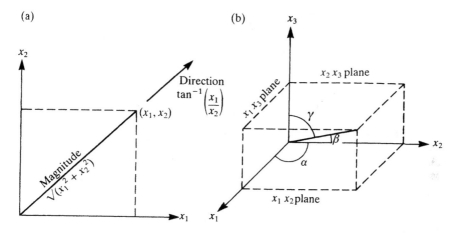

Figure 5.1 Vectors in the plane and in space. (a) Graphical representation of the vector $(x_1\ x_2)$ in the x–plane, showing magnitude and direction. (b) Graphical representation of the vector $(x_1\ x_2\ x_3)$ in space, showing the relationship to planes x_1x_2, x_1x_3 and x_2x_3, in terms of the magnitude and direction of the position vector.

$$\text{Magnitude } (L) = \sqrt{(x_1^2 + x_2^2 + x_3^2)}$$

and the direction will be defined in terms of the three axes and are calculated using the following:

$$\cos \alpha = \frac{x_1}{L}; \ \cos \beta = \frac{x_2}{L}; \ \cos \gamma = \frac{x_3}{L}$$

where L is the magnitude of the vector. This is illustrated in Fig. 5.1 b.

Example 5.2
What are the magnitudes and directions of the position vectors (1 1 1) and (3 2 1)?

Vector (1 1 1):

$$L = \sqrt{1^2 + 1^2 + 1^2}$$
$$= 1.73205$$

since $x_1 = x_2 = x_3$

then:
$$\cos \alpha = \cos \beta = \cos \gamma$$
$$= \frac{1}{1.73205}$$
$$= 0.57735$$
$$\alpha = \beta = \gamma = 54° \ 44'$$

Vector (3 2 1)

$$L = \sqrt{3^2 + 2^2 + 1^2}$$
$$= \sqrt{14}$$
$$= 3.742$$

$$\cos \alpha = \frac{3}{3.742}; \quad \cos \beta = \frac{2}{3.742}; \quad \cos \gamma = \frac{1}{3.742}$$
$$= 0.8018 \qquad\qquad = 0.5345 \qquad\qquad = 0.2673$$
$$\alpha = 36° 42' \qquad\qquad \beta = 57° 41' \qquad\qquad \gamma = 74° 29'$$

5.1.2 The angle between two vectors

Example 5.3

Consider the case of the zone axis and the crystal faces it contains. In Example 1.15 the zone axis symbol $[1, -2, 2]$ was shown to contain the faces $(2\ 1\ 0)$ and $(0\ 1\ 1)$. The dot products of the zone axis and the faces are:

$$(1\ -2\ 2) . (2\ 1\ 0) = (1 * 2) + (-2 * 1) + (2 * 0)$$
$$= 0$$

and
$$(1\ -2\ 2) . (0\ 1\ 1) = (1 * 0) + (-2 * 1) + (2*1)$$
$$= 0$$

Now by definition Miller indices of the planes $(h\ k\ l)$ of a zone are at right angles to the direction of the zone axis $[U,\ V,\ W]$ i.e. they are mutually orthogonal, which suggests that if two vectors lie at right angles to one another their dot product is zero. Indeed the usual expression used to relate the Miller indices of the plane and the directional indices of the zone axis is:

$$hU + kV + lW = 0$$

(Windle, 1977). Let us consider two further examples.

Example 5.4

Find the dot product of the following pairs of vectors and identify pairs which are mutually orthogonal:

(a) $x_1 = 1 ; x_2 = 0$
$x_1 = 0 ; x_2 = 1$

(b) $x_1 = 1 ; x_2 = 2$
$x_1 = -1 ; x_2 = 0.5$

(c) $x_1 = 1.5 ; x_2 = 3.6$
$x_1 = 2.1 ; x_2 = -1.0$

(a) $(1\ 0).(0\ 1) = (1 * 0) + (0 * 1) = 0$

(b) $(1\ 2).(-1\ 0.5) = (1 * -1) + (2 * 0.5) = 0$

(c) $(1.5 \ 3.6).(2.1 \ -1.0) = (1.5 * 2.1) + (3.6 * -1.0) = -0.45$

From the coordinates given in Example 5.4(a), it is clear that they represent vectors which are coincident with the two axes and are therefore mutually orthogonal by definition. The vectors for (b) and (c) are shown on Fig. 5.2, where it can be seen that the pair $(1 \ 2) \ (-1 \ 0.5)$ are at right angles, whereas the pair $(1.5 \ 3.6) \ (2.1 \ -1.0)$ are not.

Example 5.5

Find the dot product of the following pairs of Miller indices:

(a) Two faces of a cube which have a common edge:

$$(1 \ 0 \ 0) . (0 \ 1 \ 0) = 0$$

(b) Two faces of a rhombdodecahedron which have a common edge:

$$(1 \ 0 \ 1) . (1 \ 1 \ 0) = 1$$

In the case of the cube any pair of adjacent faces are at right angles whereas in the rhombdodecahedron they are not. The interfacial angle between $(1 \ 0 \ 1)$ and $(1 \ 1 \ 0)$ is in fact 120°.

From these examples, we are now in a position to state, without formal proof.

The dot product of two mutually orthogonal vectors is always zero.

In certain situations it is useful to be able to determine whether two linearly independent vectors **a** and **b** with a common origin define a

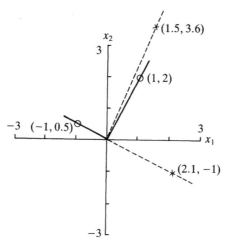

Figure 5.2 Orthogonal vectors. From the diagram it is clear that the vector pair $(1 \ 2)$ and $(-1.0 \ 0.5)$ are orthogonal (see Example 5.4b), whereas the pair $(1.5 \ 3.6)$ and $(2.1 \ -1.0)$ are not (Example 5.4c).

positively or negatively oriented dihedral angle, without necessarily calculating the angle itself. By convention a **positive dihedral angle** represents an **anti-clockwise rotation** and a **negative dihedral angle** represents a **clockwise rotation** (Fig. 5.3). As we shall see later, this rule will prove useful when transformations are considered.

In order to determine the orientation of the angle between two vectors, a matrix is formed from the vectors and its determinant calculated. Thus if the vector **a** is represented by the coordinate pair $(a_1 \ a_2)$ and vector **b** is represented by $(b_1 \ b_2)$, then the matrix is:

$$\begin{bmatrix} a_1 & a_2 \\ b_1 & b_2 \end{bmatrix}$$

and the rule is:

If the determinant of the matrix formed by writing the first vector as the first row of the matrix and the second vector as the second row is positive, then the dihedral angle is positive; whereas if it is negative, then the dihedral angle is likewise negative.

A simple example is now demonstrated by way of illustration.

Example 5.6

If the following vectors have a common origin with the vector (5.5 1.5), do they represent clockwise or anti-clockwise rotations?
The vectors are: (a) (3 4)
                    (b) (2.5 −2)
                    (c) (−1 2.5)

(a) $\det \begin{bmatrix} 5.5 & 1.5 \\ 3 & 4 \end{bmatrix} = (5.5 \times 4) - (1.5 \times 3)$

$$= 22 - 4.5 = 17.5$$

As the dihedral angle is positive the rotation is anti-clockwise.

(b) $\det \begin{bmatrix} 5.5 & 1.5 \\ 2.5 & -2 \end{bmatrix} = -11 - 3.75 = -14.75$

Therefore since the dihedral angle is negative the rotation is clockwise.

(c) $\det \begin{bmatrix} 5.5 & 1.5 \\ -1 & 2.5 \end{bmatrix} = 13.75 + 1.5 = 15.25$

Therefore since the dihedral angle is positive the rotation is anti-clockwise.
These three results are shown diagrammatically in Fig. 5.3.

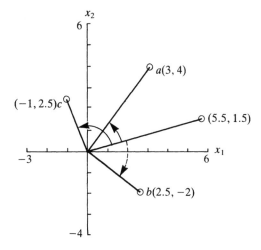

Figure 5.3 Vector rotation. Illustrating Example 5.6, demonstrating clockwise and anti-clockwise rotations from an initial vector drawn from the origin to the point (5.5, 1.5).

5.1.3 The concurrence of three straight lines

In Chapter 1 it was demonstrated that if two straight lines were defined by their slope and intercept with reference to a single pair of mutually orthogonal axes, there were a number of conditions which determined whether the lines would meet at a point without resorting to the solution of the SLE set. A problem which frequently arises is where there are three straight lines and we require to know if they are concurrent (meet at a single point). Indeed there were three examples in Chapter 1 where this problem arose, on that occasion the SLE were solved pair-wise to determine concurrence. This was clumsy and an alternative approach will now be demonstrated.

Let us suppose that there are three straight lines whose slopes are a_1, a_2 and a_3 and intercepts z_1, z_2 and z_3 relative to a pair of common axes x_1 and x_2. The equations corresponding to the lines are:

$$x_2 = a_1 x_1 + z_1 \text{ or } a_1 x_1 - x_2 + z_1 = 0 \tag{1}$$
$$x_2 = a_2 x_1 + z_2 \text{ or } a_2 x_1 - x_2 + z_2 = 0 \tag{2}$$
$$x_2 = a_3 x_1 + z_3 \text{ or } a_3 x_1 - x_2 + z_3 = 0 \tag{3}$$

Taking (1) and (2) and using Cramer's rule, the points of intersection x_{1p} and x_{2p} are:

$$x_{1p} = \frac{\begin{vmatrix} -1 & z_1 \\ -1 & z_2 \end{vmatrix}}{\begin{vmatrix} a_1 & -1 \\ a_2 & -1 \end{vmatrix}} \text{ and } -x_{2p} = \frac{\begin{vmatrix} a_1 & z_1 \\ a_2 & z_2 \end{vmatrix}}{\begin{vmatrix} a_1 & -1 \\ a_2 & -1 \end{vmatrix}}$$

If the line represented by (3) also passes through the point (x_{1p}, x_{2p}) then:

$$a_3 x_{1p} - x_{2p} + z_3 = 0$$

which means:

$$a_3 \begin{vmatrix} -1 & z_1 \\ -1 & z_2 \end{vmatrix} + \begin{vmatrix} a_1 & z_1 \\ a_2 & z_2 \end{vmatrix} + z_3 \begin{vmatrix} a_1 & -1 \\ a_2 & -1 \end{vmatrix} = 0 \tag{4}$$

also

$$a_3 \begin{vmatrix} 1 & z_1 \\ 1 & z_2 \end{vmatrix} - \begin{vmatrix} a_1 & z_1 \\ a_2 & z_2 \end{vmatrix} + z_3 \begin{vmatrix} a_1 & 1 \\ a_2 & 1 \end{vmatrix} = 0 \tag{5}$$

Remembering the method of cofactors outlined earlier (section 3.1.1), for the calculation of a 3×3 determinant, the determinantal equations (4) and (5) represent:

$$\begin{vmatrix} a_3 & 1 & z_3 \\ a_1 & 1 & z_1 \\ a_2 & 1 & z_2 \end{vmatrix} = 0 = \begin{vmatrix} a_1 & -1 & z_1 \\ a_2 & -1 & z_2 \\ a_3 & -1 & z_3 \end{vmatrix}$$

In other words if we rewrite the equations as homogeneous SLE, then

the determinant of the matrix of coefficients will be zero if the three lines are concurrent.

Example 5.7
Using the equations given in Chapter 1, Examples 1.8 and 1.9, determine whether the lines are concurrent.

(a) $\begin{aligned} 2x_1 - x_2 &= 2 \\ 6x_1 - 2x_2 &= 9 \\ -3x_1 - 8x_2 &= 13 \end{aligned}$ or $\begin{aligned} 2x_1 - x_2 - 2 &= 0 \\ 6x_1 - 2x_2 - 9 &= 0 \\ -3x_1 - 8x_2 - 13 &= 0 \end{aligned}$

We evaluate:

$$\begin{vmatrix} 2 & -1 & -2 \\ 6 & -2 & -9 \\ -3 & -8 & -13 \end{vmatrix} = 2 \begin{vmatrix} -2 & -9 \\ -8 & -13 \end{vmatrix} + 1 \begin{vmatrix} 6 & -9 \\ -3 & -13 \end{vmatrix} - 2 \begin{vmatrix} 6 & -2 \\ -3 & -8 \end{vmatrix}$$

$$= -92 - 105 + 108$$

$$= -89$$

Therefore the lines are not concurrent.

(b) $\begin{aligned} x_1 + x_2 &= 0.65 \\ 2x_1 &= 0.26 \\ 2x_2 &= 1.04 \end{aligned}$ or $\begin{aligned} x_1 + x_2 - 0.65 &= 0 \\ 2x_1 - 0.26 &= 0 \\ 2x_2 - 1.04 &= 0 \end{aligned}$

We evaluate:

$$\begin{vmatrix} 1 & 1 & -0.65 \\ 2 & 0 & -0.26 \\ 0 & 2 & -1.04 \end{vmatrix} = 1 \begin{vmatrix} 0 & -0.26 \\ 2 & -1.04 \end{vmatrix} - 1 \begin{vmatrix} 2 & -0.26 \\ 0 & -1.04 \end{vmatrix} - 0.65 \begin{vmatrix} 2 & 0 \\ 0 & 2 \end{vmatrix}$$

$$= 0.52 + 2.08 - 2.60$$
$$= 0$$

Therefore the lines are concurrent.

5.1.4 Relationship between three points

We now consider three points P_1, P_2 and P_3 in the xy–plane. If they are defined by coordinates (x_1, y_1); (x_2, y_2) and (x_3, y_3), then it can be shown that they will all lie on the same straight line i.e. they are **colinear**, if the determinant of the matrix formed from their coordinates as below, is zero:

$$\det \begin{bmatrix} x_1 & y_1 & 1 \\ x_2 & y_2 & 1 \\ x_3 & y_3 & 1 \end{bmatrix} = 0$$

Similarly it is possible to show that if three points P_1, P_2 and P_3, lying in the xyz–plane are defined by their coordinates (x_1, y_1, z_1); (x_2, y_2, z_2) and (x_3, y_3, z_3), will lie in the **same plane**, if:

$$\det \begin{bmatrix} x_1 & y_1 & z_1 \\ x_2 & y_2 & z_2 \\ x_3 & y_3 & z_3 \end{bmatrix} = 0$$

Examples of the use of these two relationships are given in the Student examples below.

5.1.5 Student examples

Q.5.1   Calculate the magnitude and direction of the following vectors:

(a) (1 2)   (b) (5 −8)
(c) (4 6)   (d) (−3 2)

Q5.2   (a) The kilometre (3-figure) grid references of three town centres are:

Newcastle-upon-Tyne	424 564
Morpeth	420 585
Hexham	393 564

Calculate the distances between the towns and determine the directions of travel going from Newcastle-upon-Tyne to Morpeth, Morpeth to Hexham and Hexham to Newcastle-upon-Tyne.

(b) Would a journey from Morpeth to Newcastle represent a clockwise or an anti-clockwise rotation?

Q5.3   Calculate the magnitude and directions of the vectors:

$$\text{(a) (2 1 4)} \qquad \text{(b) (1 2 } \tfrac{1}{2}\text{)}$$

Q5.4   Which of the following pairs of vectors are mutually orthogonal?

(a) (0 1) (1 0)   (b) (2 3) (4 2)
(c) (2 3 −1) (2 −1 3)   (d) (1 −1 −1) (−1 −1 0)
(e) (1 0 −1 1) (1 −1 0 −1)   (f) (2 0 −2 3) (1 0 −1 2)
(g) (2 0 −2 3) (2 0 −1 −2)   (h) (2 1 −3 1) (0 1 1 2)

Q5.5   During a survey three traverses are made, their origins and directions are:

$$\text{(1) (1.4, 0) } 35° ; \quad \text{(2) (0, 1.5) } 47° ; \quad \text{(3) (3.54, 6.5) } 67°$$

Do the three traverses meet at a single point? If so, where is it? Hint: use tangents to 2 dp and calculate intercepts on *y*–axis (northings).

Q5.6   Three boreholes cut through the top of a limestone unit at 30 m, 35 m and 40 m respectively. If their coordinates are (3, 7), (7.5, 6.5) and (3, 3), does the limestone form a plane surface?

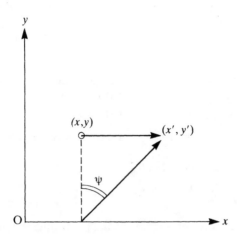

Figure 5.4 Simple shear in 2-dimensions. The point (*x*, *y*) is displaced parallel to the *x*–axis to a new position (*x'*, *y'*). The angle ψ is the angular shear strain.

5.2 Transformations

Consider the case of a point (x, y) in the xy–plane (throughout this section the xy–plane will be used in line with the convention adopted in the geological literature), which is displaced parallel to the x–axis, to some new point (x', y') as in Fig. 5.4. The vector joining the point in its initial position (x, y) with the new position (x', y') can be resolved into a component u which is parallel to the x–axis and a component v which is parallel to the y–axis:
and:

$$u = x' - x$$
$$v = y' - y$$

since $y' = y$, then the displacement equations are:

$$u = \gamma\, y$$
$$v = 0$$

where $\gamma = \tan \psi$ where ψ is the **angular shear strain** and γ (gamma) is the **shear strain**.

The coordinate transform equations are:

$$x' = x + \gamma\, y$$
$$y' = 0x + y$$

and written in matrix form:

$$\begin{bmatrix} x' \\ y' \end{bmatrix} = \begin{bmatrix} 1 & \gamma \\ 0 & 1 \end{bmatrix} \cdot \begin{bmatrix} x \\ y \end{bmatrix}$$

from which we get the **strain matrix:**

$$\begin{bmatrix} 1 & \gamma \\ 0 & 1 \end{bmatrix}$$

Note: if the displacement had been parallel to y–axis, then the strain matrix would be:

$$\begin{bmatrix} 1 & 0 \\ \gamma & 1 \end{bmatrix}$$

In practice if we multiply the coordinates of any point in the xy–plane by the strain matrix we will get a new point whose displacement will be parallel to the x-axis (or y-axis).

This is **simple shear**, which can be defined as:

A displacement which transforms an initial square into a parallelogram. The displacement vectors are parallel to one set of opposite sides. There is no area change.

In the examples of simple shear which follow we shall represent the shear strain by a positive number in the matrix. Students requiring information on the sign convention relating to simple shear should consult Ramsay and Huber (1983, Appendix A).

Another deformation known as **pure shear**, is defined as strain without rotation with no area change. The typical strain matrix for pure shear is:

$$\begin{bmatrix} k & 0 \\ 0 & \frac{1}{k} \end{bmatrix}$$

and the transform equations are:

$$x' = kx$$
$$y' = \frac{y}{k}$$

Pure shear deforms an initial square into a rectangle with both sets of displacement vectors parallel to the axes.

Many other transformation matrices are recognized, some of which have no geological application. We consider briefly three examples:

$$\mathbf{A} = \begin{bmatrix} k & 0 \\ 0 & k \end{bmatrix} \qquad \mathbf{B} = \begin{bmatrix} -k & 0 \\ 0 & k \end{bmatrix} \qquad \mathbf{C} = \begin{bmatrix} -k & 0 \\ 0 & -k \end{bmatrix}$$

The matrix $\mathbf{A}$ defines radial expansion, unless $k = 1$ in which case we have the unit matrix and there will be no change whatsoever.

The matrix $\mathbf{B}$ produces a 'mirror' image, and if the absolute value $k > 1$, then this will be accompanied by radial expansion. This transformation is not possible geologically.

The matrix $\mathbf{C}$ defines body rotation through 180° and again, if the absolute value of $k > 1$, this will be accompanied by radial expansion and is possible geologically.

We now return to the cases of pure shear and simple shear, which we shall illustrate using simple examples. To eliminate the effect of body translation in these examples, the centre of the figure to be deformed will be defined as the origin of the axes, since the point (0, 0) is always invariant under transformation.

Example 5.8

Using the unit square, illustrate the effects of pure shear and simple shear, if the constants are:

$$k = 2$$
$$\psi = 1.5$$

The square is identified by the corners A, B, C and D with coordinate pairs (x_1, y_1); (x_2, y_2); (x_3, y_3) and (x_4, y_4) locating their positions in the xy−plane. The operations are:

For pure shear –

$$\begin{bmatrix} k & 0 \\ 0 & \frac{1}{k} \end{bmatrix} \cdot \begin{bmatrix} x_n \\ y_n \end{bmatrix} = \begin{bmatrix} k * x_n \\ \frac{1}{k} * y_n \end{bmatrix} = \begin{bmatrix} x_n' \\ y_n' \end{bmatrix}$$

and for simple shear –

$$\begin{bmatrix} 1 & \gamma \\ 0 & 1 \end{bmatrix} \cdot \begin{bmatrix} x_n \\ y_n \end{bmatrix} = \begin{bmatrix} x_n + (\gamma * y_n) \\ y_n \end{bmatrix} = \begin{bmatrix} x_n' \\ y_n' \end{bmatrix}$$

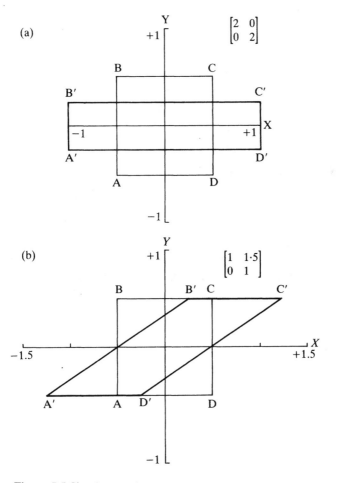

Figure 5.5 Simple transformation of the unit square ABCD, using
(a) pure and
(b) simple shear.
Transformation matrices as given in Example 5.8. The transformed points are labelled A'B'C'D'. Students are reminded that for this and other related examples, operations are carried out in the xy–plane.

The results are illustrated in Fig. 5.5a (pure shear) and Fig. 5.5b (simple shear). Notice that the deformations are as predicted by the definitions given earlier, also there are no area changes. The determinants of the two matrices can also be calculated:

$$\det \begin{bmatrix} k & 0 \\ 0 & \frac{1}{k} \end{bmatrix} = k * \frac{1}{k} = 1$$

$$\det \begin{bmatrix} 1 & \gamma \\ 0 & 1 \end{bmatrix} = (1 * 1) - (0 * \gamma) = 1$$

Example 5.9

Investigate the deformation of a unit square, by a two-stage process:

(a) pure shear followed by simple shear;
(b) simple shear followed by pure shear.

The solution can be obtained by multiplying (x_n, y_n) the coordinates of any corner, by the first shear matrix to give the transformed coordinates (x_n', y_n'), which can then be multiplied by the second matrix. Alternatively the two matrices can be multiplied together in the **reverse order** of the events and the resulting matrix used directly, to give (x_n', y_n'). We shall use the second method for both examples.

(a) For pure shear followed by simple shear:

$$\begin{bmatrix} 1 & \gamma \\ 0 & 1 \end{bmatrix} \cdot \begin{bmatrix} k & 0 \\ 0 & \frac{1}{k} \end{bmatrix} = \begin{bmatrix} k & \frac{\gamma}{k} \\ 0 & \frac{1}{k} \end{bmatrix}$$

which using the values for γ and k given in Example 5.8, gives:

$$\begin{bmatrix} 2 & 0.75 \\ 0 & 0.5 \end{bmatrix} \tag{1}$$

The result of multiplying the coordinates of the unit square by matrix (1) is shown in Fig. 5.6a.

(b) Reversing the order of the deformation events so that simple shear is followed by pure shear, we get:

$$\begin{bmatrix} k & 0 \\ 0 & \frac{1}{k} \end{bmatrix} \cdot \begin{bmatrix} 1 & \gamma \\ 0 & 1 \end{bmatrix} = \begin{bmatrix} k & k\gamma \\ 0 & \frac{1}{k} \end{bmatrix}$$

which using the values as before is:

$$\begin{bmatrix} 2 & 3 \\ 0 & 0.5 \end{bmatrix} \tag{1}$$

and the result of this operation (multiplying the coordinates by matrix (2)), is shown in Fig. 5.6b.

Notice that as a consequence of the fact that matrix multiplication does not commute, the result of the two-stage deformation process involving

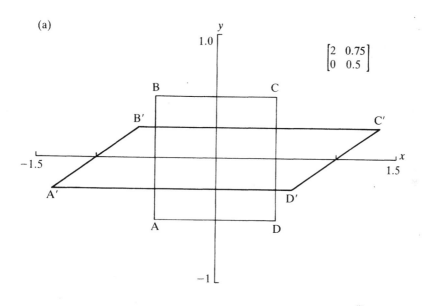

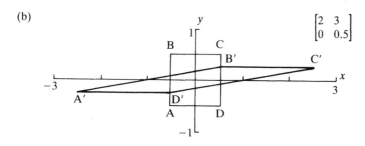

Figure 5.6 Transformation of the unit square ABCD after two transformation events: (a) pure shear followed by simple shear; and (b) simple shear followed by pure shear. Transformation matrices as given in Example 5.9.

both pure shear and simple shear is dependent on the order of the events. Thus the result in matrix algebra:

$$\mathbf{A \cdot B \neq B \cdot A}$$

has important consequences for the structural geologist, since in many situations deformation is a multi-stage process involving different styles of deformation. Thus if the order of events is not clear, it becomes difficult to interpret the resulting deformation. The student should also note that the determinants of the two matrices, (1) and (2) above, are still unity.

Turning now to the inverse problem: if the deformed shape has been observed and recorded can the original be recovered? The general equations we have already shown are:

$$x' = ax + by$$
$$y' = cx + dy$$

in matrix form these are:

$$\begin{bmatrix} x' \\ y' \end{bmatrix} = \begin{bmatrix} a & b \\ c & d \end{bmatrix} \cdot \begin{bmatrix} x \\ y \end{bmatrix} \tag{1}$$

thus given x, y we can calculate x', y'. However if we are given x', y' then this will be equivalent to solving the SLE:

$$ax + by = x'$$
$$cx + dy = y'$$

the solution of which we have already demonstrated in section 3.2.1, as:

$$\begin{bmatrix} x \\ y \end{bmatrix} = \begin{bmatrix} a & b \\ c & d \end{bmatrix}^{-1} \cdot \begin{bmatrix} x' \\ y' \end{bmatrix} \tag{2}$$

Hence to find the original coordinates given their deformed position we multiply by the inverse of the strain matrix.

The matrix equation set (1) is the **Lagrangian Specification** or strain matrix form of the transform equations, while (2) is the reciprocal strain matrix form or **Eulerian equations**.

Geologically deformation unaccompanied by area or volume change, such as in the cases of simple shear or pure shear just considered, is rare. This limitation is overcome in the concept of **general homogeneous strain**, which is defined as:

> strain where any set of parallel lines change their orientation and spacing distance but remain straight and parallel. The deformation may be accompanied by a change in area.

The situation where there is no area change is termed **plane strain**, where there is an area change it is termed **non-plane strain**. Simple shear and pure shear are special cases of general homogeneous strain and are examples of plane strain.

The coordinate transform equations for general homogeneous strain are:

$$x' = ax + by$$
$$y' = cx + dy$$

and the matrix form is:

$$\begin{bmatrix} x' \\ y' \end{bmatrix} = \begin{bmatrix} a & b \\ c & d \end{bmatrix} \cdot \begin{bmatrix} x \\ y \end{bmatrix} \tag{1}$$

We now consider examples where the elements of the strain matrix are not equal to zero.

Example 5.10
Using the unit square centred at the origin, demonstrate the deformation produced by the following strain matrices:

(a) $\begin{bmatrix} 2 & 2 \\ 2 & 4 \end{bmatrix}$ (b) $\begin{bmatrix} 2 & 4 \\ 3 & 4 \end{bmatrix}$

What are the determinants of these two matrices? The coordinates of the unit square are $(-\frac{1}{2}, -\frac{1}{2})$ $(-\frac{1}{2}, \frac{1}{2})$ $(\frac{1}{2}, \frac{1}{2})$ and $(\frac{1}{2}, -\frac{1}{2})$. The new coordinates of the four corners are calculated as follows.

(a) $\begin{bmatrix} 2 & 2 \\ 2 & 4 \end{bmatrix} \cdot \begin{bmatrix} -\frac{1}{2} \\ -\frac{1}{2} \end{bmatrix} = \begin{bmatrix} -2 \\ -3 \end{bmatrix}$ (b) $\begin{bmatrix} 2 & 4 \\ 3 & 4 \end{bmatrix} \cdot \begin{bmatrix} -\frac{1}{2} \\ -\frac{1}{2} \end{bmatrix} = \begin{bmatrix} -3 \\ -3\frac{1}{2} \end{bmatrix}$

$\cdot \begin{bmatrix} -\frac{1}{2} \\ \frac{1}{2} \end{bmatrix} = \begin{bmatrix} 0 \\ 1 \end{bmatrix}$ $\cdot \begin{bmatrix} -\frac{1}{2} \\ \frac{1}{2} \end{bmatrix} = \begin{bmatrix} 1 \\ \frac{1}{2} \end{bmatrix}$

$\cdot \begin{bmatrix} \frac{1}{2} \\ \frac{1}{2} \end{bmatrix} = \begin{bmatrix} 2 \\ 3 \end{bmatrix}$ $\cdot \begin{bmatrix} \frac{1}{2} \\ \frac{1}{2} \end{bmatrix} = \begin{bmatrix} 3 \\ 3\frac{1}{2} \end{bmatrix}$

$\cdot \begin{bmatrix} \frac{1}{2} \\ -\frac{1}{2} \end{bmatrix} = \begin{bmatrix} 0 \\ -1 \end{bmatrix}$ $\cdot \begin{bmatrix} \frac{1}{2} \\ -\frac{1}{2} \end{bmatrix} = \begin{bmatrix} -1 \\ -\frac{1}{2} \end{bmatrix}$

and the determinants of the two matrices:

(a)   Det $\begin{bmatrix} 2 & 2 \\ 2 & 4 \end{bmatrix} = 4$ (b)   Det $\begin{bmatrix} 2 & 4 \\ 3 & 4 \end{bmatrix} = -4$

The results of the two transformations are illustrated in Fig. 5.7, examination of which can help elucidate some principles. Clearly both transformations demonstrate that original parallel lines are still parallel although their length and orientation have changed, also there is a change in area in both cases, factors which are all accounted for by the definition given earlier. An additional consequence of the second transformation (Example 5.10 (b)), is the production of a 'mirror' image. This as noted earlier, is not possible geologically.

Returning to the results for simple and pure shear (Examples 5.8 and 5.9), the determinants of the matrices used were unity, whereas the determinants of the matrices in Example 5.10 have values of 4.0 and −4.0. Inspection of Fig. 5.7 shows that for both transformations there is an increase in area. From these and earlier observations, it is apparent that there is a relationship between the determinant of the strain matrix and the area change of the transformed figure. Before drawing any firm conclusions we will consider two further examples.

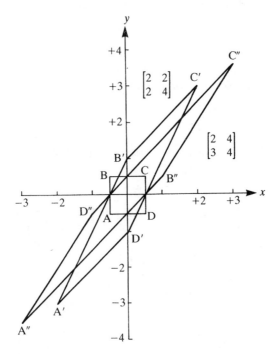

Figure 5.7 The transformation of the unit square ABCD, under general homogeneous strain. Matrices as given in Example 5.10. Notice that the transformed figure A'B'C'D' shows shear accompanied by an increase in area, whereas for the transformation leading to A"B"C"D", these effects are also accompanied by reflection about A"C", the points B→B" and D→D" being reversed.

Example 5.11
Using the unit square, demonstrate the deformation produced by the following strain matrices:

$$\text{(a)} \begin{bmatrix} 1 & 0.4 \\ 1 & 0.5 \end{bmatrix} \qquad \text{(b)} \begin{bmatrix} 3 & 1.5 \\ 4 & 2 \end{bmatrix}$$

What are the determinants of the matrices and what is the area relationship between the original and transformed figures?

$$\text{(a)} \begin{bmatrix} 1 & 0.4 \\ 1 & 0.5 \end{bmatrix} \cdot \begin{bmatrix} 1 \\ 1 \end{bmatrix} = \begin{bmatrix} 1.4 \\ 1.5 \end{bmatrix} \qquad \text{(b)} \begin{bmatrix} 3 & 1.5 \\ 4 & 2 \end{bmatrix} \cdot \begin{bmatrix} 1 \\ 1 \end{bmatrix} = \begin{bmatrix} 4.5 \\ 6 \end{bmatrix}$$

$$\cdot \begin{bmatrix} 1 \\ 2 \end{bmatrix} = \begin{bmatrix} 1.8 \\ 2 \end{bmatrix} \qquad\qquad \cdot \begin{bmatrix} 1 \\ 2 \end{bmatrix} = \begin{bmatrix} 6 \\ 8 \end{bmatrix}$$

$$\cdot \begin{bmatrix} 2 \\ 2 \end{bmatrix} = \begin{bmatrix} 2.8 \\ 3 \end{bmatrix} \qquad\qquad \cdot \begin{bmatrix} 2 \\ 2 \end{bmatrix} = \begin{bmatrix} 9 \\ 12 \end{bmatrix}$$

$$\cdot \ \begin{bmatrix} 2 \\ 1 \end{bmatrix} = \begin{bmatrix} 2.4 \\ 2.5 \end{bmatrix} \qquad\qquad \cdot \ \begin{bmatrix} 2 \\ 1 \end{bmatrix} = \begin{bmatrix} 7.5 \\ 10 \end{bmatrix}$$

$$\text{Det} \ \begin{bmatrix} 1 & 0.4 \\ 1 & 0.5 \end{bmatrix} = 0.1 \qquad \text{Det} \ \begin{bmatrix} 3 & 1.5 \\ 4 & 2 \end{bmatrix} = 0$$

The result of the transformation represented by the matrix (a) is shown in Fig. 5.8, inspection of which shows that the transformed figure is much smaller than the original. The results for matrix (b), show that there is a constant ratio of (3:4) between each pair of x', y' coordinate values which implies that they plot as a straight line. Thus we may conclude that if the determinant of the strain matrix is zero, one dimension is lost. The results which have been obtained from the various examples used in this section can now be summarized as in Table 5.1

Inspection of Table 5.1 leads to the conclusion that there is a simple relationship between the area change or **dilation** (normally represented by the Greek capital letter 'delta' Δ_A) and the determinant of the matrix, thus:

$$1 + \Delta_A = (ad - bc)$$
$$= \text{determinant of the strain matrix}$$

(Ramsay and Huber, 1983, p.287). Since two of the examples considered are not possible geologically: i.e. the loss of one dimension (determinant zero), and that producing a 'mirror' image (determinant negative), we are restricted in practical geological situations to matrices whose determinants

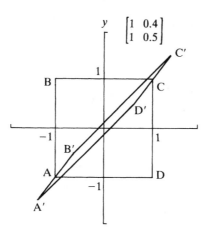

Figure 5.8 The transformation of the unit square ABCD, under general homogenous strain, Matrix as given in Example 5.11a. Notice that the transformed figure A'B'C'D' shows shear, accompanied by a decrease in area but with no change of sense.

Table 5.1 Summary of the results from Examples 5.8 to 5.11, relating the area of the transformed figure and the determinant of the transformation matrix

Value of determinant	Result of deformation
0	Loss of one dimension
1	No area change
>1	Increase in area
>0 <1	Decrease in area
<0	'Mirror' image + change in area

are greater than zero. From the stand-point of exploration whether for minerals or for oil or gas, deformations which are accompanied by changes in volume are important, since this could create or destroy the pore space required to accommodate a material of economic importance.

Finally before leaving this section we consider an example which demonstrates homogeneous strain in 3–dimensions.

Example 5.12

Using a unit cube with centre $(0,0,0)$, demonstrate the deformation produced by the matrix:

$$\begin{bmatrix} 1 & 1 & 2 \\ 2 & 1 & 1 \\ 1 & 2 & 1 \end{bmatrix}$$

In this example, each corner of the cube will be defined with reference to a vector in the xyz–plane. Thus if the position of the back bottom right hand corner is $(-0.5, 0.5, -0.5)$ then the new position (x', y', z'), will be:

$$\begin{bmatrix} 1 & 1 & 2 \\ 2 & 1 & 1 \\ 1 & 2 & 1 \end{bmatrix} \cdot \begin{bmatrix} -0.5 \\ 0.5 \\ -0.5 \end{bmatrix} = \begin{bmatrix} -1.0 \\ -1.0 \\ 0 \end{bmatrix}$$

The other seven corners of the cube are defined accordingly and their new positions calculated. The transformed figure is illustrated in Fig. 5.9, which demonstrates that the original cube has been subjected to homogeneous shear, it has been rotated and the deformation has been accompanied by an increase in volume. The dilation is:

$$1 + \Delta_V = \det \begin{bmatrix} 1 & 1 & 2 \\ 2 & 1 & 1 \\ 1 & 2 & 1 \end{bmatrix} = 4$$

where Δ_v is the dilation or change in volume.

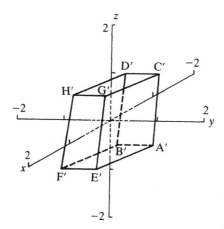

Figure 5.9 A partial representation of the transformation of a unit cube, by the matrix given in Example 5.12. On the scale represented in the diagram, the transformed figure shows an increase in volume.

Before leaving this section we should emphasize once again the fact that although we have largely illustrated transformations in 2–dimensions, the results and conclusions are equally valid in 3–dimensions.

5.2.1 Progressive deformation

In the examples of deformation considered in the previous section, the process has been treated as a single step and we have been concerned simply with the comparison of the original and final states. More realistically and taking into consideration that geological events are more often than not incremental, perhaps taking place over very long periods of time, consideration should be given to progressive changes. Thus an original set of points is displaced in a series of discrete steps, simulated by multiplying the coordinate set of the original or previous deformation the requisite number of times.

Example 5.13
Using the matrices and initial points given, plot the deformation paths generated during 4 discrete time steps:

(a) Pure shear where $k = 1.25$; initial point (5, 5)
(b) Simple shear where $\gamma = 0.7$; initial point (1, 1)
(c) General homogeneous strain defined by the matrix:

$$\begin{bmatrix} 1.1 & 0.5 \\ 0.2 & 1.2 \end{bmatrix}$$

and initial point (1, 1).

(a) For pure shear, the deformation matrix is:

$$\begin{bmatrix} 1.25 & 0 \\ 0 & 0.8 \end{bmatrix}$$

using t_0 to indicate the initial point, the coordinates in vector form are:

$$t_0 = \begin{bmatrix} 5 \\ 5 \end{bmatrix} \quad t_1 = \begin{bmatrix} 6.25 \\ 4 \end{bmatrix} \quad t_2 = \begin{bmatrix} 7.813 \\ 3.2 \end{bmatrix} \quad t_3 = \begin{bmatrix} 9.766 \\ 2.56 \end{bmatrix} \quad t_4 = \begin{bmatrix} 12.207 \\ 2.048 \end{bmatrix}$$

(b) For simple shear, the deformation matrix is:

$$\begin{bmatrix} 1 & 0.7 \\ 0 & 1 \end{bmatrix}$$

and the coordinates are:

$$t_0 = \begin{bmatrix} 1 \\ 1 \end{bmatrix} \quad t_1 = \begin{bmatrix} 1.7 \\ 1 \end{bmatrix} \quad t_2 = \begin{bmatrix} 2.4 \\ 1 \end{bmatrix} \quad t_3 = \begin{bmatrix} 3.1 \\ 1 \end{bmatrix} \quad t_4 = \begin{bmatrix} 3.8 \\ 1 \end{bmatrix}$$

(c) For general homogeneous strain, the deformation matrix is:

$$\begin{bmatrix} 1.1 & 0.5 \\ 0.3 & 1.2 \end{bmatrix}$$

and the coordinates are:

$$t_0 = \begin{bmatrix} 1 \\ 1 \end{bmatrix} \quad t_1 = \begin{bmatrix} 1.6 \\ 1.5 \end{bmatrix} \quad t_2 = \begin{bmatrix} 2.51 \\ 2.28 \end{bmatrix} \quad t_3 = \begin{bmatrix} 3.901 \\ 3.489 \end{bmatrix} \quad t_4 = \begin{bmatrix} 6.0356 \\ 5.3571 \end{bmatrix}$$

The results are plotted as Fig. 5.10 which shows the new positions of the points for each time increment, for the three deformations.

The observant reader may have noticed that if we had been asked to find only the final point after four time increments, then the same result would have been given by:

$$[A]^4 \cdot \begin{bmatrix} x \\ y \end{bmatrix}_{t_0} = \begin{bmatrix} x' \\ y' \end{bmatrix}_{t_4}$$

where A is the strain matrix and the vectors as defined earlier. Taking Example 5.13 (a), we have:

$$\begin{bmatrix} 1.25 & 0 \\ 0 & 0.8 \end{bmatrix}^4 \cdot \begin{bmatrix} 5 \\ 5 \end{bmatrix} = \begin{bmatrix} x' \\ y' \end{bmatrix}_{t_4}$$

$$\begin{bmatrix} 2.442 & 0 \\ 0 & 0.41 \end{bmatrix} \cdot \begin{bmatrix} 5 \\ 5 \end{bmatrix} = \begin{bmatrix} 12.21 \\ 2.05 \end{bmatrix}$$

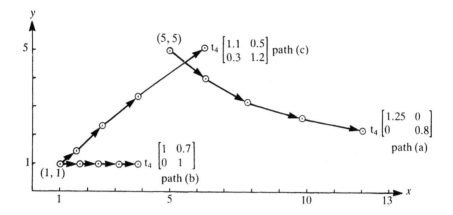

Figure 5.10 A plot of the transformation paths of the point (1, 1) under simple shear, general homogeneous strain (two examples) and the point (5, 5) under pure shear; all for 4 increments of deformation, as Example 5.13.

The small differences between x', y' here and those given earlier are simply due to rounding-off during powering the matrix.

Example 5.13
Using the general homogeneous strain matrix in Example 5.13 (c), show the progressive deformation during three time increments, of a rectangle whose coordinates in vector form are:

$$\begin{bmatrix} 0 \\ 0 \end{bmatrix} \begin{bmatrix} 0 \\ 4 \end{bmatrix} \begin{bmatrix} 2 \\ 4 \end{bmatrix} \begin{bmatrix} 2 \\ 0 \end{bmatrix}$$

The coordinates of the four corners are:

$$t_0 = \begin{bmatrix} 0 \\ 0 \end{bmatrix}, \quad t_1 = \begin{bmatrix} 0 \\ 0 \end{bmatrix}, \quad t_2 = \begin{bmatrix} 0 \\ 0 \end{bmatrix}, \quad t_3 = \begin{bmatrix} 0 \\ 0 \end{bmatrix}$$

$$t_0 = \begin{bmatrix} 2 \\ 0 \end{bmatrix}, \quad t_1 = \begin{bmatrix} 2.2 \\ 0.6 \end{bmatrix}, \quad t_2 = \begin{bmatrix} 2.72 \\ 1.38 \end{bmatrix}, \quad t_3 = \begin{bmatrix} 3.68 \\ 2.47 \end{bmatrix}$$

$$t_0 = \begin{bmatrix} 0 \\ 4 \end{bmatrix}, \quad t_1 = \begin{bmatrix} 2.0 \\ 4.8 \end{bmatrix}, \quad t_2 = \begin{bmatrix} 4.6 \\ 6.36 \end{bmatrix}, \quad t_3 = \begin{bmatrix} 8.24 \\ 9.01 \end{bmatrix}$$

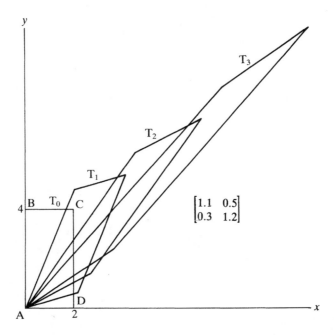

Figure 5.11 A plot showing the transformation of the rectangle ABCD under general homogeneous strain for 3 increments of deformation, as Example 5.14.

$$t_0 = \begin{bmatrix} 2 \\ 4 \end{bmatrix}, \quad t_1 = \begin{bmatrix} 4.2 \\ 5.4 \end{bmatrix}, \quad t_2 = \begin{bmatrix} 7.32 \\ 7.74 \end{bmatrix}, \quad t_3 = \begin{bmatrix} 11.92 \\ 11.48 \end{bmatrix}$$

Figure 5.11 shows this data plotted graphically.

5.2.2 Student examples

Q5.7 Transform a unit square, coordinates (0, 0); (0, 1); (1, 1) and (1, 0) by:
  (a) pure shear, where $k = 1.25$;
  (b) simple shear, where $\gamma = 1.4$; and
  (c) using the same constants find the result of simple shear followed by pure shear.

Q5.8 The transformed positions (x', y') of 4 points after general homogeneous strain are: (2.9, 3); (3.9, 5); (4.8, 4) and (5.8, 6).

If the strain matrix was

$$\begin{bmatrix} 1.9 & 1 \\ 2 & 2 \end{bmatrix}$$

what were the original positions?

Q5.9 Using the strain matrix

$$\begin{bmatrix} 2 & 1 \\ 1 & 2 \end{bmatrix}$$

show that the progressive deformation after 4 steps is the same as the deformation produced by the fourth power of the matrix.

Q5.10 Show that the result:

$$(\mathbf{B} \cdot \mathbf{A}^3) \cdot \begin{bmatrix} x \\ y \end{bmatrix} = \begin{bmatrix} x' \\ y' \end{bmatrix}$$

where **A** is some matrix representing simple shear
  B is some matrix representing pure shear
is the same as 3 steps of simple shear (with γ constant) followed by pure shear, when:

(a) $(x, y) = (2.5, 3)$; $\gamma = 0.105$; $k = 1.1$
(b) $(x, y) = (4, 2)$; $\gamma = 0.09$; $k = 1.2$

Q5.11 Draw graphs showing the following transformations:

(a) Matrix $\begin{bmatrix} 2 & 2 \\ 2 & 4 \end{bmatrix}$ coordinates $\begin{bmatrix} 2 \\ 0.5 \end{bmatrix}$ and $\begin{bmatrix} 1 \\ 1.618 \end{bmatrix}$

(b) Matrix $\begin{bmatrix} 2 & 4 \\ 3 & 4 \end{bmatrix}$ coordinates $\begin{bmatrix} 2 \\ 0.5 \end{bmatrix}$ and $\begin{bmatrix} 1 \\ 1.151 \end{bmatrix}$

Using the method described earlier (section 5.1.4) test for colinearity between the origin (0, 0) the original point and the transformed point, for each of the four points given.

Q5.12 Selecting suitable points on the circumference of a unit circle with centre (0, 0), calculate the new positions after deformation under homogeneous strain defined by the matrix:

$$\begin{bmatrix} 2 & 2 \\ 2 & 4 \end{bmatrix}$$

Sketch the result, can you identify the new shape?
  Hint: Use Pythagoras' theorem to calculate the coordinates required. If there are gaps after your initial selection of points, select others to fill them, so that your sketch is reasonably accurate.

Starting at the point (0, 0) draw in the long axis of the new figure, from this using $x = 3$ determine y. Multiply this pair of coordinates by the matrix and plot this new point on your sketch. What is the relationship between these pairs of coordinates?

IMPORTANT – keep your answers to Q5.11 and Q5.12. You will find them useful when reading the next chapter.

6

Eigenvalues and eigenvectors

6.1 What are eigenvalues and eigenvectors?

The student examples Q5.11 and Q5.12, both involved the multiplication of pairs of coordinates by a strain matrix. The results of these operations are illustrated as Figs 6.1 a and b referring to Q5.11 and Fig. 6.2 referring to Q5.12. The figures you have drawn should be similar to those illustrated here. It should be noticed that in the case of Q5.11 (both parts), a straight line drawn from the origin passing through the original coordinates also passes through the transformed coordinates, for one of each of the given coordinate pairs only. This fact should have been confirmed by the test for colinearity, which it was suggested, should be used to check the results.

The observant student will also have noticed that in Q5.12, when:

$$x = 3 \text{ and } y \simeq 4.875$$

the transformed coordinates also lie on the straight line from the origin i.e. the points (0, 0); (3, 4.875) and (15.75, 25.5) are colinear. Again this can be confirmed by using the test for colinearity on the three coordinate pairs, when the determinant of the matrix will be found to be approximately equal to zero (small discrepancies are due to inaccuracies in reading values from a diagram). In Q5.11 the coordinate pair (2, 0.5) and their transforms are not colinear with the origin, as shown in Fig. 6.1 a and b.

Experimentation with the figure produced in answer to Q5.12 will confirm that with two exceptions (which includes the point already discussed), points chosen at random on the circumference of the unit circle with centre (0, 0) are not colinear with their transforms and the origin.

We could demonstrate by trial and error that for any given 2×2 matrix there are only two vectors which have this property of colinearity. These vectors are special since these and only these, show pure stretch with no rotation. Before they can be studied in detail, a method must be developed to enable their orientation to be found, other than by trial and error. Vectors which have this property are known as **eigenvectors** and associated with each vector is a numerical value known as the **eigenvalue**, which defines the length of the individual vector.

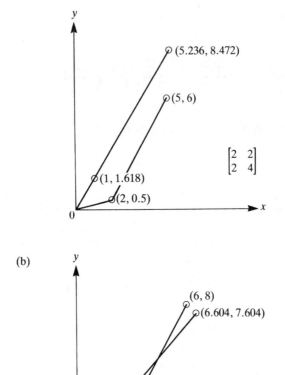

Figure 6.1 Graphs showing the answers to Student example Q5.11: (a) Q5.11a; and (b) Q5.11b.

We start with a more formal definition of eigenvalues and eigenvectors:

Let **A** be a square matrix of size n with real components. The number λ (lambda) which can be real or complex, is called an eigenvalue of **A** if there exists a nonzero vector **v** so that

$$\mathbf{A}.\mathbf{v} = \lambda.\mathbf{v}$$

The vector $\mathbf{v} \neq \mathbf{0}$ is called an eigenvector of the matrix **A** corresponding to the eigenvalue λ (The vectors which we have just been discussing whose slope was not changed by the transformation, are the eigenvectors **v**).

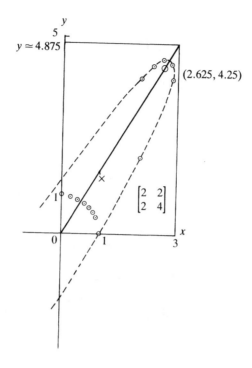

Figure 6.2 Graph showing the answer to Student example Q5.12. Notice that because of scaling the diagram, the point labelled (2.625, 4.25) is a scaled version of the product of the matrix and the coordinate pair found when $x = 3$.

If $\mathbf{A} = \mathbf{I}$,

then $\mathbf{A.v} = \mathbf{I.v} = \mathbf{v}$

and $\lambda = 1$

However if λ is an eigenvalue of $\mathbf{A}$ then there is a nonzero vector:

$$\mathbf{v} = \begin{bmatrix} v_1 \\ v_2 \\ . \\ . \\ v_n \end{bmatrix} \neq \mathbf{0}$$

such that:

$$\mathbf{A.v} = \lambda\mathbf{v} = \lambda\mathbf{I.v}$$

which can be rewritten:

$$(\mathbf{A} - \lambda\mathbf{I})\mathbf{v} = \mathbf{0}$$

If $\mathbf{A}$ is an $n \times n$ matrix this equation represents a homogeneous system of n equations with unknowns $x_1, x_2, \cdots, x_n$. If $\det(\mathbf{A} - \lambda\mathbf{I}) = 0$ then the equation has nontrivial solutions and λ is an eigenvalue of $\mathbf{A}$. If on the other hand $\det(\mathbf{A} - \lambda\mathbf{I}) \neq 0$, then the equation has only the solution $\mathbf{v} = \mathbf{0}$ so that λ is not an eigenvalue of $\mathbf{A}$.

Remember – this only applies to square matrices, rectangular matrices do not have eigenvectors.

To summarize we need to find values of λ which satisfy the equation:

$$\det(\mathbf{A} - \lambda\mathbf{I}) = 0$$

which will allow us to find values $v_1, v_2, \cdots, v_n$ the elements of the vector $\mathbf{v} \neq \mathbf{0}$. This equation is known as the **characteristic equation** of the matrix $\mathbf{A}$.

6.2 Calculation of eigenvalues and eigenvectors

6.2.1 Solving the characteristic equation

At this stage, rather than using the general case, we consider the situation where $\mathbf{A}$ is a 2×2 matrix. Thus:

$$\mathbf{A} = \begin{bmatrix} a & b \\ c & d \end{bmatrix}$$

and

$$\lambda\mathbf{I} = \begin{bmatrix} \lambda & 0 \\ 0 & \lambda \end{bmatrix} \qquad \text{(Note: matrix same order as } \mathbf{A})$$

therefore:

$$(\mathbf{A} - \lambda\mathbf{I}) = \begin{bmatrix} a & b \\ c & d \end{bmatrix} - \begin{bmatrix} \lambda & 0 \\ 0 & \lambda \end{bmatrix} = \begin{bmatrix} a - \lambda & b \\ c & d - \lambda \end{bmatrix}$$

Now we need to find values of λ, such that:

$$\det \begin{bmatrix} a - \lambda & b \\ c & d - \lambda \end{bmatrix} = 0$$

We expand the determinant:

$$\det = ((a - \lambda)(d - \lambda)) - (c * b)$$
$$= \lambda^2 - (a + d)\lambda + (ad - bc)$$

Expanding the determinant of any 2×2 matrix gives a quadratic equation. In general if

$$y = ax^2 + bx + c$$

then the roots of the equation will give values for x when $y = 0$. Similarly for the equation:

$$\det = \lambda^2 - (a + d)\lambda + (ad - bc)$$

the roots will give values for λ when $\det = 0$.

If the roots are not distinct then the eigenvectors will be coincident.

In the general case if A is an $n \times n$ matrix then the order of the expanded determinant will also be n. By the fundamental theorem of algebra a polynomial equation of degree n, will have exactly n roots (real or complex), counting multiplicities. Thus it can said that since any eigenvalue of A is a root of the equation found by expanding the determinant of A, then counting multiplicities every $n \times n$ matrix has n eigenvalues. Returning to the 2×2 case there will be two eigenvalues λ_1 and λ_2, which can be real and equal, real and distinct or complex. The roots can be put into the original equations to allow the calculation of values x_1 and y_1 for the eigenvector v_1, associated with λ_1 and x_2 and y_2 for the eigenvector v_2, associated with λ_2. The equations are:

$$(a - \lambda_1)x_1 + by_1 = 0$$
$$cx_1 + (d - \lambda_1)y_1 = 0$$

and:

$$(a - \lambda_2)x_2 + by_2 = 0$$
$$cx_2 + (d - \lambda_2)y_2 = 0$$

giving two eigenvectors:

$$v_{\lambda_1} = \begin{bmatrix} x_1 \\ y_1 \end{bmatrix} \quad ; \quad v_{\lambda_2} = \begin{bmatrix} x_2 \\ y_2 \end{bmatrix}$$

We should also note the very important fact:

If the matrix A is symmetrical, i.e. $A = A^T$ then the eigenvectors will be mutually orthogonal. In all other cases they will not.

Example 6.1
Find the eigenvalues and eigenvectors of the matrix used in student examples Q5.11a and Q5.12.
The matrix is:

$$\begin{bmatrix} 2 & 2 \\ 2 & 4 \end{bmatrix}$$

Step 1 Write out the characteristic equation:

$$\det \begin{bmatrix} [A] - [\lambda I] \end{bmatrix} = \det \begin{bmatrix} 2 - \lambda & 2 \\ 2 & 4 - \lambda \end{bmatrix} = 0$$

Step 2 Expand the determinant:

$$((2 - \lambda)(4 - \lambda)) - (2 * 2) = 0$$
$$(\lambda^2 - 6\lambda + 8) - 4 = 0$$
$$\lambda^2 - 6\lambda + 4 = 0$$

Step 3   Solve the quadratic equation, to give the eigenvalues:

$$\lambda = \frac{6 \pm \sqrt{36 - 16}}{2}$$

$$\lambda = 5.236 \text{ or } 0.764$$

Step 4   Using the two eigenvalues calculated, solve the homogeneous equations, to give the associated eigenvectors:

$$\left(2 - \lambda_n\right) x_n + 2y_n \quad = 0$$
$$2x_n + \left(4 - \lambda_n\right) y_n \quad = 0$$

For $\lambda_1 = 5.236$

$$(2 - 5.236)x_1 + 2y_1 \quad = 0$$
$$2x_1 + (4 - 5.236)y_1 = 0$$

which is:

$$-3.236x_1 + 2y_1 \quad = 0$$
$$2x_1 + -1.236y_1 = 0$$

if $x_1 = 1$
then:

$$-1.236y_1 = -2$$
$$y_1 = 1.618$$

thus:

$$\mathbf{v}_{\lambda_1} = \begin{bmatrix} 1 \\ 1.618 \end{bmatrix}$$

For $\lambda_2 = 0.764$

$$(2 - 0.764)x_2 + 2y_2 \quad = 0$$
$$2x_2 + (4 - 0.764)y_2 = 0$$

which is:

$$1.236x_2 + 2y_2 \quad = 0$$
$$2x_2 + 3.236y_2 = 0$$

if $x_2 = 1$
then:

$$3.236y_2 = -2$$
$$y_2 = -0.618$$

thus:

$$\mathbf{v}_{\lambda_2} = \begin{bmatrix} 1 \\ -0.618 \end{bmatrix}$$

Since the matrix we have used is symmetrical we can expect that the vectors will be orthogonal. We test to ensure that the eigenvectors are orthogonal, to confirm this expectation:

$$[1 \ 1.618] \cdot [1 \ -0.618] = (1 - 0.9999)$$
$$= 0$$

Returning for a moment to the solution to student example Q5.12, where it asked that the long axis of the transformed figure be drawn in. If its length from the origin to the point at which it cuts the circumference of the ellipse is now measured, it will be found to be approximately 5.25 units. This length is near enough equal to the largest eigenvalue of the matrix to allow the suggestion that: starting with a circle of unit radius with centre (0 0), the transformation produced by a symmetrical strain matrix, is an ellipse whose major semi-axis has the same unit length as the largest eigenvalue of the strain matrix. Also the direction of the major semi-axis will be given by the corresponding eigenvector. Referring to Fig. 6.2 we can see that the point labelled X, on the axis of the ellipse is (1, 1.62) and vector ox is the eigenvector of the transformation matrix used.

As we shall see later if the strain matrix is not symmetrical, then the major and minor semi-axes of the ellipse will not be coincident with the eigenvectors.

Also note that in general:

(a) The determinant of a matrix is equal to the product of its eigenvalues.

Hence for the matrix under consideration:

$$\lambda_1 * \lambda_2 = \det \begin{bmatrix} 2 & 2 \\ 2 & 4 \end{bmatrix} = 5.236 * 0.764$$
$$= 4.0$$

(b) The sum of the elements of the principal diagonal of the matrix is equal to the sum of the eigenvalues.

For the last example:

$$a_{1,1} + a_{2,2} = 6$$
$$\lambda_1 + \lambda_2 = 6$$

From the characteristic equation it is clear that if the eigenvalues are back substituted into the matrix, then the determinants of the resulting matrices will equal zero. This can be used to check the calculations and for this example we have:

$$\det \begin{bmatrix} (2 - 5.236) & 2 \\ 2 & (4 - 5.236) \end{bmatrix} = \det \begin{bmatrix} -3.236 & 2 \\ 2 & -1.236 \end{bmatrix} = 4 - 4 = 0$$

and:

$$\det \begin{bmatrix} (2 - 0.764) & 2 \\ 2 & (4 - 0.764) \end{bmatrix} = \det \begin{bmatrix} 1.236 & 2 \\ 2 & 3.236 \end{bmatrix} = 4 - 4 = 0$$

The method outlined, based on expanding the determinant, can be extended to deal with a matrix of any size. Since the order of the polynomial generated will be the same as the size of the matrix, difficulties arise in practice when finding its roots. Let us consider the case of a 3×3 matrix.

Example 6.2
Expand the determinant of the matrix, to give its characteristic polynomial:

$$\begin{bmatrix} 1 - \lambda & 1 & 2 \\ 2 & 1 - \lambda & 1 \\ 1 & 2 & 1 - \lambda \end{bmatrix}$$

That is:

$$(1 - \lambda) \begin{vmatrix} 1 - \lambda & 1 \\ 2 & 1 - \lambda \end{vmatrix} - \begin{vmatrix} 2 & 1 \\ 1 & 1 - \lambda \end{vmatrix} + 2 \begin{vmatrix} 2 & 1 - \lambda \\ 1 & 2 \end{vmatrix}$$

Taking each minor separately:

(i) $(1 - \lambda) \begin{vmatrix} 1 - \lambda & 1 \\ 2 & 1 - \lambda \end{vmatrix} = -\lambda^3 + 3\lambda^2 - \lambda - 1$

(ii) $- \begin{vmatrix} 2 & 1 \\ 1 & 1 - \lambda \end{vmatrix} = -(2 - 2\lambda - 1)$

(iii) $2 \begin{vmatrix} 2 & 1 - \lambda \\ 1 & 2 \end{vmatrix} = 2(4 - (1 - \lambda))$
$= (6 + 2\lambda)$

Collecting terms gives $-\lambda^3 + 3\lambda^2 + 3\lambda + 4$. Thus the characteristic equation is

$$-\lambda^3 + 3\lambda^2 + 3\lambda + 4 = 0 \tag{1}$$

If we wish to proceed further and calculate the eigenvalues of the original 3×3 matrix, the roots of the cubic equation (1) above, will have to be found. And, although this particular equation appears innocuous, finding its roots is not easy. Thus for matrices larger than 2×2, or for specially constructed examples, other methods of solution must be found. This problem will be looked at in the next section, when a simple iterative process will be considered.

Finally, if the matrix is in diagonal form – all elements zero except those in the principal diagonal (for example the matrix describing pure shear), then the eigenvalues will have the same values as the elements in the diagonal. Also, the eigenvectors will be coincident with the axes. If the matrix is in upper or lower triangular form – all elements below or above the principal diagonal are zero with values elsewhere (for example the matrix describing simple shear as used in Example 5.7 is in upper triangular form), then the eigenvalues will also have the same values as the elements in the diagonal. In the simple 2×2 case, the eigenvectors will be coincident with one of the axes – with the x–axis if the matrix is in upper triangular form, or with the y–axis for a lower triangular form.

6.2.2 Student examples

Q6.1   Find the eigenvalues and eigenvectors of the matrices:

$$\text{(a)} \quad \begin{bmatrix} 2 & 4 \\ 3 & 4 \end{bmatrix} \qquad \text{(b)} \quad \begin{bmatrix} 1 & 0.4 \\ 1 & 0.5 \end{bmatrix}$$

Show that the eigenvectors are not mutually orthogonal.

Q6.2   Expand the determinant of:

$$\begin{bmatrix} 1.9 & 1 & 1 \\ 1 & 2 & 1 \\ 1 & 1 & 1.9 \end{bmatrix} - \lambda\,[\mathbf{I}]$$

Q6.3   Using the matrices for simple and pure shear given in Example 5.7, expand the determinants and hence calculate the eigenvalues and eigenvectors, demonstrating the truth of the statements in the last paragraph of section 6.2.1.

Q6.4   Find the eigenvalues and eigenvectors of the matrix:

$$\begin{bmatrix} 1.9 & 1 \\ 1 & 2 \end{bmatrix}$$

Keep your results for use with the discussion in section 6.3.

6.2.3 An iterative method to find eigenvalues and eigenvectors

If some matrix $[\mathbf{A}]$ has n linearly independent eigenvectors and a dominant eigenvalue λ_1 (i.e. $\lambda_1 > \lambda_i$ for $i = 2, \dots , n$), then:

$$\alpha_j^{(k+1)} = \frac{j\text{th component } \mathbf{A}^{k+1}.\mathbf{x}_0}{j\text{th component } \mathbf{A}^k.\mathbf{x}_0} = \lambda_1$$

where $\mathbf{x}_0$ is some arbitrarily chosen vector with the same number of rows as the matrix has columns, and:

$$\mathbf{A}^k.\mathbf{x}_0 = \mathbf{x}_k$$

where $\mathbf{x}_k$ is an eigenvector corresponding to the eigenvalue λ_1. This is known as the **power method**. This method will only work if there is a dominant eigenvalue. The mathematics of the method is best understood with reference to an example.

Example 6.3
Use the power method to find the dominant eigenvalue and the corresponding eigenvector of the matrix:

$$\begin{bmatrix} 1.9 & 1 \\ 1 & 2 \end{bmatrix}$$

Step 1   Choose the vector x_0:
   Since the vector x_0 is chosen arbitrarily it is convenient to choose a simple value such as:

$$x_0 = (1\ 1)$$

Step 2   Commence the iteration by calculating x_1 and the ratio $\alpha_1^{(k+1)}$ and $\alpha_2^{(k+1)}$:

$$x_1 = A.x_0 = \begin{bmatrix} 1.9 & 1 \\ 1 & 2 \end{bmatrix} \cdot \begin{bmatrix} 1 \\ 1 \end{bmatrix} = \begin{bmatrix} 2.9 \\ 3 \end{bmatrix}$$

$$\alpha_1^{(1)} = \frac{2.9}{1} \qquad \alpha_2^{(1)} = \frac{3}{1}$$

$$= 2.9 \qquad\qquad = 3$$

Step 3   Continue the iteration until $\alpha_1 = \alpha_2$, to some pre-determined degree of accuracy.

$$x_2 = A.x_1 = \begin{bmatrix} 1.9 & 1 \\ 1 & 2 \end{bmatrix} \cdot \begin{bmatrix} 2.9 \\ 3 \end{bmatrix} = \begin{bmatrix} 8.51 \\ 8.9 \end{bmatrix}$$

$$\alpha_1^{(2)} = \frac{8.51}{2.9} \qquad \alpha_2^{(2)} = \frac{8.9}{3}$$

$$= 2.9345 \qquad\qquad = 2.9667$$

At this point it is convenient to tabulate the results of further iterations as in Table 6.1.

Table 6.1 Tabulated results for the solution of Example 6.3

Iteration	x_k (as a row vector)		$\alpha_1^{(k)}$	$\alpha_2^{(k)}$
0	(1	1)	–	–
1	(2.9	3)	2.9	3
2	(8.51	8.9)	2.9345	2.9667
3	(25.069	26.31)	2.9458	2.9562
4	(73.9411	77.689)	2.9495	2.9528
5	(218.177	229.3191)	2.95069	2.95175
6	(643.8556	676.8153)	2.95107	2.95141

From the table of results it is clear that λ_{max} for this matrix is 2.951 (correct to 3 dp) and that $v_{\lambda_{max}}$ (the eigenvector) will be:

$$(1 \ \tfrac{676.8153}{643.8556}) = (1 \ 1.051)$$

Those who have attempted student example Q6.4 by expanding the determinant, will be able to verify that these figures are very close to the true values which (correct to 4 dp) are:

$$\lambda_{max} = 2.9513 \text{ and } v_{\lambda_{max}} = (1 \ 1.0526)$$

A significant problem which should be immediately apparent from this example is the rate of growth of the size of the numerical values for the vector x_k. This problem can be overcome by normalizing the vector x_k by dividing through by its largest component.

Example 6.4
Re-do Example 6.3 using the power method with scaling.

Step 1   Choose the vector x_0:

$$x_0 = (1 \ 1)$$

Step 2   Commence the iteration by calculating x_1:

$$x_1 = A.x_0 = \begin{bmatrix} 1.9 & 1 \\ 1 & 2 \end{bmatrix} . \begin{bmatrix} 1 \\ 1 \end{bmatrix} = \begin{bmatrix} 2.9 \\ 3 \end{bmatrix}$$

Step 3   Normalize the vector x_1:

$$x'_1 = \frac{1}{3} \begin{bmatrix} 2.9 \\ 3 \end{bmatrix} = \begin{bmatrix} 0.9667 \\ 1 \end{bmatrix}$$

Step 3   Continue the iteration to find x_2:

$$x_2 = A.x'_1 = \begin{bmatrix} 1.9 & 1 \\ 1 & 2 \end{bmatrix} . \begin{bmatrix} 0.9667 \\ 1 \end{bmatrix} = \begin{bmatrix} 2.8367 \\ 2.9667 \end{bmatrix}$$

Step 4   Normalize the vector x_2:

$$x'_2 = \frac{1}{2.9667} \begin{bmatrix} 2.8367 \\ 2.9667 \end{bmatrix} = \begin{bmatrix} 0.95619 \\ 1 \end{bmatrix}$$

Remember – since the eigenvalue we are trying to find is always the largest element of x_k, then its value before normalization is the approximation to the eigenvalue. In this example element x_2 is the largest, thus at Step 3 the approximation to λ_{max} is 2.9667.

Step 5   Continue iterations until the difference between successive values of x_k reaches some predetermined amount, dependent on the degree of

Table 6.2 Tabulated results for the power method with scaling,
Example 6.4

Iteration	$\mathbf{x}_k$	$\mathbf{x}'_k$ (normalized)
0	(1 1)	(1 1)
1	(2.9 3)	(0.9667 1)
2	(2.83673 2.9667)	(0.95619 1)
3	(2.816762 2.95619)	(0.9528 1)
4	(2.810386 2.9528)	(0.95177 1)
5	(2.808363 2.95177)	(0.95142 1)
6	(2.807698 2.95142)	(0.95130 1)

accuracy required. At this point it is again convenient to tabulate the results, these are shown in Table 6.2.

It is clear from the table that the results:

$$\lambda_{max} = 2.951$$

and

$$\mathbf{v}_{\lambda_{max}} = (1 \quad \frac{1}{0.9513})$$

$$= (1 \quad 1.0512)$$

are as we had before.

The disadvantage of the method described is that the largest eigenvalue only, is found. In order to calculate further eigenvalues we use the following result.

If λ_1 , λ_2 , $\cdots$, λ_n are the eigenvalues of the matrix $\mathbf{A}_1$ and λ_1 is the dominant eigenvalue with eigenvector $\mathbf{v}_1$ then, let $\mathbf{u}$ be a column vector such that:

$$\mathbf{v}_1 \cdot \mathbf{u} = 1$$

If the matrix $\mathbf{A}_2$ is given by:

$$\mathbf{A}_2 = \mathbf{A}_1 - \lambda_1 * (\mathbf{v}_1 \cdot \mathbf{u}^T)$$

then the eigenvalues of $\mathbf{A}_2$ are 0 , λ_2 , λ_3 , $\cdots$, λ_n and λ_2 , λ_3 , $\cdots$, λ_n are also eigenvalues of matrix $\mathbf{A}_1$.

Remember – the result of multiplying two column vectors together is a scalar. However, if the second vector is transposed (i.e. it is then a row vector), multiplication will result in a matrix of order n, where n is the number of rows in the original vectors (Chapter 2, section 2.2.3).

Now if the matrix $\mathbf{A}_2$ has a dominant eigenvalue we can use the power method with scaling to find that eigenvalue and its associated eigenvector. These will also be the second eigenvalue and associated eigenvector for the original matrix $\mathbf{A}_1$. Thus by repeating the calculation with a suitable choice of vectors **u** we can find all of the eigenvalues of matrix A_1, provided $\lambda_1 > \lambda_2 > \cdots > \lambda_n$. The method is known as the **power method with deflation** (scaling can be incorporated as necessary).

Example 6.5

Find the eigenvalues and eigenvectors of the matrix:

$$\begin{bmatrix} 1.9 & 1 & 1 \\ 1 & 2 & 1 \\ 1 & 1 & 1.9 \end{bmatrix}$$

using the power method with deflation and scaling.

In the step by step outline of the method which follows, details relating to the calculation of the eigenvalues and eigenvectors are omitted for clarity.

Step 1   Use power method with scaling to find the largest eigenvalue and associated eigenvector. These are:

Eigenvalue (1): $\lambda_1 = 3.934$

Eigenvector (1):
$$\mathbf{v}_{\lambda_1} = \begin{bmatrix} 1 \\ 1 \\ 1 \end{bmatrix}$$

Step 2   Find a column vector **u** which when the scalar product is taken with $\mathbf{v}_{\lambda_1}$, gives a value of unity:

$$\mathbf{v}_{\lambda_1} \cdot \mathbf{u} = 1$$

A suitable choice would be:

$$\begin{bmatrix} 1 \\ 1 \\ 1 \end{bmatrix} \cdot \begin{bmatrix} 1/3 \\ 1/3 \\ 1/3 \end{bmatrix} = 1/3 + 1/3 + 1/3 = 1$$

The calculations:

$$\mathbf{A}_2 = \mathbf{A}_1 - \lambda_1 * (\mathbf{v}_{\lambda_1} \cdot \mathbf{u}^T)$$

are now carried out in 2 steps:

Step 3   Calculate $\lambda_1.(\mathbf{v}_{\lambda_1} \cdot \mathbf{u}^T)$:

$$\lambda_1 \cdot \begin{bmatrix} 1 \\ 1 \\ 1 \end{bmatrix} (\tfrac{1}{3} \ \tfrac{1}{3} \ \tfrac{1}{3}) = 3.934 \cdot \begin{bmatrix} \tfrac{1}{3} & \tfrac{1}{3} & \tfrac{1}{3} \\ \tfrac{1}{3} & \tfrac{1}{3} & \tfrac{1}{3} \\ \tfrac{1}{3} & \tfrac{1}{3} & \tfrac{1}{3} \end{bmatrix}$$

$$= \begin{bmatrix} 1.3113 & 1.3113 & 1.3113 \\ 1.3113 & 1.3113 & 1.3113 \\ 1.3113 & 1.3113 & 1.3113 \end{bmatrix}$$

Step 4   Subtract the resulting matrix from $\mathbf{A}_1$, to give the new matrix $\mathbf{A}_2$:

$$\mathbf{A}_2 = \begin{bmatrix} 1.9 & 1 & 1 \\ 1 & 2 & 1 \\ 1 & 1 & 1.9 \end{bmatrix} - \begin{bmatrix} 1.3113 & 1.3113 & 1.3113 \\ 1.3113 & 1.3113 & 1.3113 \\ 1.3113 & 1.3113 & 1.3113 \end{bmatrix}$$

$$= \begin{bmatrix} 0.5887 & -0.3113 & -0.3113 \\ -0.3113 & 0.6887 & -0.3113 \\ -0.3113 & -0.3113 & 0.5887 \end{bmatrix}$$

Step 5   Use the power method with scaling to find the largest eigenvalue and the associated eigenvector for matrix $\mathbf{A}_2$.

These were found to be:

Eigenvalue (2): $\lambda_2 = 0.969$

Eigenvector (2):

$$\mathbf{v}_{\lambda_2} = \begin{bmatrix} -0.4501 \\ 1 \\ -0.4501 \end{bmatrix}$$

Step 6   Repeat steps 2, 3, 4 and 5:

Step 2:

$$\begin{bmatrix} -0.4501 \\ 1 \\ -0.4501 \end{bmatrix} \cdot \begin{bmatrix} -1 \\ 1 \\ 1 \end{bmatrix} = 0.4501 + 1 - 0.4501 = 1$$

Step 3:

$$\lambda_2 \cdot \begin{bmatrix} -0.4501 \\ 1 \\ -0.4501 \end{bmatrix} \cdot (-1 \ \ 1 \ \ 1) = 0.969 \cdot \begin{bmatrix} 0.4501 & -0.4501 & -0.4501 \\ -1 & 1 & 1 \\ 0.4501 & -0.4501 & -0.4501 \end{bmatrix}$$

$$= \begin{bmatrix} 0.43615 & -0.43615 & -0.43615 \\ -0.969 & 0.969 & 0.969 \\ 0.43615 & -0.43615 & -0.43615 \end{bmatrix}$$

Step 4: Remember we use matrix $\mathbf{A}_2$ in this operation.

$$\mathbf{A}_3 = \begin{bmatrix} 0.5887 & -0.3113 & -0.3113 \\ -0.3113 & 0.6887 & -0.3113 \\ -0.3113 & -0.3113 & 0.5887 \end{bmatrix} - \begin{bmatrix} 0.43615 & -0.43615 & -0.43615 \\ -0.969 & 0.969 & 0.969 \\ 0.43615 & -0.43615 & -0.43615 \end{bmatrix}$$

$$= \begin{bmatrix} 0.15255 & 0.12485 & 0.12485 \\ 0.6577 & -0.2803 & -1.2803 \\ -0.74745 & 0.12485 & 1.02485 \end{bmatrix}$$

Step 5: Use the power method with scaling to find the largest eigenvalue and the associated eigenvector for matrix $\mathbf{A}_3$.

These were found to be:

Eigenvalue (3) $\lambda_3 = 0.867$

Eigenvector (3)

$$\mathbf{v}_{\lambda_3} = \begin{bmatrix} -0.020 \\ 1 \\ -0.886 \end{bmatrix}$$

It is left to students to check that the results obtained for this example are correct (use either method given in section 6.2.1).

Remember, as with all iterative methods, the results of a particular solution are dependent on earlier calculations. As can be seen this is of particular importance in this method, where at Step 3 the eigenvalue calculated previously is used in the formation of a new matrix which is then used as the basis of the calculation of the next eigenvector and its associated eigenvalue. The choice of the vector **u** at any particular stage is arbitrary, an incorrect choice usually leads to impossible values. For example when finding λ_3 in the above example, the vector:

$$\begin{bmatrix} -0.22 \\ 1 \\ 0.22 \end{bmatrix}$$

was used. This led to the result: $\lambda_3 = 1.786$
since this value is greater than that calculated for λ_2, it is not a possible result.

There are a variety of other methods to enable eigenvalue and eigenvector problems to be solved, and most University computer centres have programmes which enable solutions to be obtained in different circumstances. Except in simple cases, students are advised to seek help and advice on methods which are available.

6.2.4 Student examples

Q6.5 Use the power method with scaling to calculate the dominant eigenvalue and its associated eigenvector for the matrix:

$$\begin{bmatrix} 2 & 4 \\ 1 & 3 \end{bmatrix}$$

correct to 4 dp.

Q6.6 Use the power method with scaling to calculate the dominant eigenvalue and its associated eigenvector for the matrix:

$$\begin{bmatrix} 1 & 1 & 2 \\ 2 & 1 & 1 \\ 1 & 2 & 1 \end{bmatrix}$$

Q6.7 Use the power method with scaling to calculate the dominant eigenvalue and its associated eigenvector for the matrix:

$$\begin{bmatrix} 4 & 2 & 3 \\ 2 & 0 & 2 \\ 2 & 2 & 4 \end{bmatrix}$$

6.3 The finite strain ellipse – a geometrical interpretation

The finite strain ellipse arises from the multiplication of the coordinates of points on the circumference of a circle of unit radius with centre $(0, 0)$, by a strain matrix. This is an important concept which has allowed scientists to standardize the illustration and description of the results of different types of strain.

To demonstrate the concept, Fig. 6.3 shows the transformation of the unit circle for the two special cases of homogeneous strain, namely simple shear (Fig. 6.3a), pure shear (Fig. 6.3b) as well as a more general example (Fig. 6.3c). The matrices used are taken from Examples 5.7 and 5.9a, Chapter 5 respectively, and are:

$$\text{A.} \begin{bmatrix} 1 & 1.5 \\ 0 & 1 \end{bmatrix} \quad \text{B.} \begin{bmatrix} 2 & 0 \\ 0 & 0.5 \end{bmatrix} \quad \text{C.} \begin{bmatrix} 2 & 2 \\ 2 & 4 \end{bmatrix}$$

The result of each deformation is an ellipse, known as the **strain ellipse**. The two mutually orthogonal semi-axes of the strain ellipse marked on the figures, are where the extensions of radii of the original circles achieve maximum and minimum values, these are the **principal strains**. The lengths of the semi-axes are $(1 + e_1)$ and $(1 + e_2)$, where e_1 and e_2 are the **principal extensions**.

For the matrices **A** and **B** we can use the definitions given in section 6.2.1 to define the numerical values for the eigenvalues and eigenvectors:

For the simple shear matrix there will be one eigenvalue $= 1.0$ (roots coincident), with a single eigenvector coincident with the x–axis.

For pure shear the eigenvalues will be 2.0 and 0.5, the first with an eigenvector coincident with the x–axis and the second coincident with the y–axis.

The values for the eigenvalues and eigenvectors for the third matrix, are (Example 6.1):

$$\lambda_1 = 5.236 \qquad \lambda_2 = 0.764$$
$$v_{\lambda_1} = (1 \ 1.618) \qquad v_{\lambda_2} = (1 \ -0.618)$$

Using Fig. 6.3, the length and direction of the major and minor semi-axes of the ellipse for each deformation can be measured and the values determined compared with those for the eigenvalues and eigenvectors of the matrices which produced the transformations as noted above. This data is summarized in Table 6.3.

From the data given in Table 6.3, it is clear that for pure shear and for the general example (**C**), the lengths and directions of the semi-axis of the

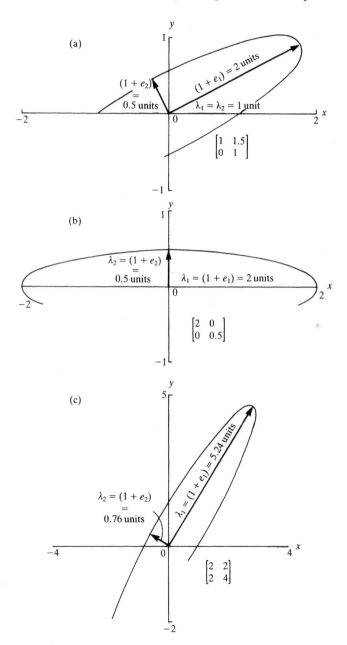

Figure 6.3 Transformation of the unit circle, centre (0, 0) under: (a) simple shear; (b) pure shear; and (c) general homogeneous strain.
Each part of the figure shows the length and the direction of the principal axes of the strain ellipse produced by the transformation illustrated, as well as the eigenvectors (see text for details).

strain ellipse are defined by the eigenvalues and eigenvectors of the strain matrices. In the case of simple shear the single eigenvector (coincident with the x-axis), is **not** coincident with either of the axes of the strain ellipse and the single eigenvalue does not equal the length of either semi-axis. The only immediately obvious difference which might account for these facts is that the simple shear matrix is not symmetrical about the principal diagonal, whereas the other two are.

In order to investigate these apparent associations further, and since it is impossible to measure lengths and angles of the semi-axes of the strain ellipse accurately, we need to be able to calculate their values. Suitable equations are given by Ramsay and Huber (1983, p.64), these are:

$$(1 + e_1)^2 = \tfrac{1}{2}(a^2 + b^2 + c^2 + d^2 + ((a^2 + b^2 + c^2 + d^2)^2 \\ - 4(ad - bc)^2)^{\frac{1}{2}})$$

$$(1 + e_2)^2 = \tfrac{1}{2}(a^2 + b^2 + c^2 + d^2 - ((a^2 + b^2 + c^2 + d^2)^2 \\ - 4(ad - bc)^2)^{\frac{1}{2}})$$

where $(1 + e_1)$ is the length of the major semi-axis
   $(1 + e_2)$ is the length of the minor semi-axis
   a, b, c and d are the elements of the strain matrix $\begin{bmatrix} a & b \\ c & d \end{bmatrix}$

$$\tan 2\theta' = \frac{2(ac + bd)}{a^2 + b^2 - c^2 - d^2}$$

$$\tan 2\theta = \frac{2(ab + cd)}{a^2 - b^2 + c^2 - d^2}$$

where $\tan \theta'$ is the slope of one of the semi-axes of the strain ellipse after deformation, the other will be at right angles. Tan θ is the slope of the vector before deformation – equivalent to multiplying the vector (1 tan θ') by the inverse of the strain matrix.

Note: – negative angles are measured clockwise from the positive x–axis, positive angles are anticlockwise.

The difference between the direction of the axis before and after rotation, is given by:

$$\tan \omega = \frac{c - b}{a + d}$$

This last equation obeys the rules for dihedral angles – if tan ω is negative then the rotation is clockwise, if it is positive then the rotation is anticlockwise.

By putting the values for the 3 matrices used to construct Fig. 6.3 into these equations, we can check accurately the conclusions reached. These results are summarized, along with values for the eigenvalues and the slopes represented by the corresponding eigenvectors, in Table 6.4.

The results obtained by calculation (Table 6.4), confirm the earlier conclusions. Additionally, they show that for simple shear there is a

Table 6.3 Comparison of eigenvalues/eigenvectors for the strain matrices A, B and C and the lengths and direction of the semi-axes of the strain ellipses measured using Fig. 6.3. All values given to 2 dp.

Deformation	λ_1		λ_2		Major semi-axis		Minor semi-axis	
	Length	Slope	Length	Slope	Length	Slope	Length	Slope
Simple Shear **A**	1	0			2	0.5	0.5	−2
Pure Shear **B**	2	0	0.5	−∞	2	0	0.5	−∞
Example **C**	5.24	1.62	0.76	−0.62	5.25	1.6	0.75	−0.62

Table 6.4 Calculated values for the principal extensions, slopes, rotations, eigenvalues and corresponding slopes of the eigenvectors for the matrices used to construct Fig. 6.3. All values correct to 4 dp.
NOTE – slopes are measured anticlockwise from the horizontal x–axis.

For simple shear, matrix: $\begin{bmatrix} 1 & 1.5 \\ 0 & 1 \end{bmatrix}$	$(1 + e_1) = 2.0$ $(1 + e_2) = 0.5$ $\tan \theta' = 0.5$ slope $= 26° 30'$ rotation $= -36° 54'$	$\lambda_1 = 1$ $\lambda_2 = 1$ slope $v_\lambda = 0$
For pure shear, matrix: $\begin{bmatrix} 2 & 0 \\ 0 & 0.5 \end{bmatrix}$	$(1 + e_1) = 2$ $(1 + e_2) = 0.5$ $\tan \theta' = 0$ rotation $= 0°$	$\lambda_1 = 2$ $\lambda_2 = 0.5$ slope $v_{\lambda_1} = 0$ slope $v_{\lambda_2} = -\infty$ mutually orthogonal
General example, matrix: $\begin{bmatrix} 2 & 2 \\ 2 & 4 \end{bmatrix}$	$(1 + e_1) = 5.2361$ $(1 + e_2) = 0.7639$ $\tan \theta' = -0.618$ slope $= 58° 17'$ rotation $= 0°$	$\lambda_1 = 5.2361$ $\lambda_2 = 0.7639$ slope $v_{\lambda_1} = 1.618$ slope $v_{\lambda_2} = -0.618$ mutually orthogonal

Table 6.5 Calculated values for the principal extensions, slopes, rotations, eigenvalues and corresponding slopes of the eigenvectors for the given matrix. All values to 4 dp.

Matrix: $\begin{bmatrix} 1 & 0.6 \\ 0.4 & 1.5 \end{bmatrix}$	$(1 + e_1) = 1.8130$ $(1 + e_2) = 0.6950$ $\tan \theta' = -0.6746$ slope $= 34°$ rotation $= -4° 35'$	$\lambda_1 = 1.8$ $\lambda_2 = 0.7$ slope $v_{\lambda_1} = 1.3333$ slope $v_{\lambda_2} = -0.5$ not mutually orthogonal

rotation of the principal axis of the strain ellipse. We will examine a further example, where the strain matrix is not symmetrical about the principal diagonal. The matrix chosen for this demonstration is:

$$\begin{bmatrix} 1 & 0.6 \\ 0.4 & 1.5 \end{bmatrix}$$

and the results obtained are again tabulated (Table 6.5).

From the results of this example given in Table 6.5, three important points are immediately apparent:

1. The eigenvalues are only approximately equal to the principal extensions.
2. The slopes of the eigenvectors are not equal to the slopes of the principal strains.
3. There is a rotational component which alters the original directions of the principal strains after transformation. This feature was also seen earlier in the case of simple shear and again, the strain matrix was not symmetrical about the principal diagonal.

The fact of rotation of the axis of principal strain, has led to the recognition of two types of general homogeneous strain:

(a) Without a rotational component, represented by the general matrix:

$$\begin{bmatrix} a & b \\ c & d \end{bmatrix} \quad (b = c)$$

with pure shear as a special case.

(b) With a rotational component, represented by the general matrix:

$$\begin{bmatrix} a & b \\ c & d \end{bmatrix} \quad (b \neq c)$$

with simple shear as a special case.

These are termed **general homogeneous irrotational strain** and **general homogeneous rotational strain** respectively.

As a final demonstration to illustrate these very important concepts consider two further examples. First we examine Example 6.4, which since the matrix is symmetrical, by definition is an example of homogeneous irrotational strain. The answers which should have been found are:

Matrix: $\begin{bmatrix} 1.9 & 1 \\ 1 & 2 \end{bmatrix}$

Eigenvalues: $\lambda_{max} = 2.95125$
$\lambda_{min} = 0.94875$

Eigenvectors $v_{\lambda_{max}} = (1\ 1.05125)$
$v_{\lambda_{min}} = (-1\ 0.95125)$

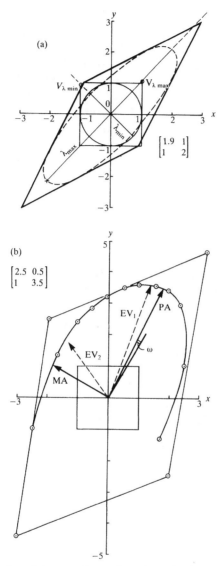

Figure 6.4 Examples of the transformation of a unit circle and a unit square, both with centre (0, 0) (see text for details): (a) demonstrating general homogeneous irrotational strain, $v_{\lambda_{max}}$ – the eigenvector for the largest eigenvalue λ_{max}, $v_{\lambda_{min}}$ – the eigenvector for the smallest eigenvalue λ_{min}; (b) demonstrating general homogeneous rotational strain and showing the relationship between the eigenvectors and the axes of the strain ellipse, as well as the angle of rotation.

EV$_1$ – largest eigenvalue PA – Principal semi-axis of the strain ellipse
EV$_2$ – smallest eigenvalue MA – Minor semi-axis of the strain ellipse
ω – rotational component.

These results are illustrated in Fig. 6.4a, which has been drawn illustrating the deformation of a unit square and a unit circle, both having a centre (0, 0), by general homogeneous irrotational strain. Examination of this figure shows that the lengths and directions of the semi-axes of the strain ellipse are equal to the calculated eigenvalues and eigenvectors of the matrix. Students may care to verify this conclusion by calculating the corresponding lengths and directions of the axes of the strain ellipse, using Ramsay and Huber's equations given earlier. The transformation also shows that there has been an area increase of approximately ×3. The determinant of the strain matrix is 2.8.

For comparison, the second example illustrating the deformation of the same square and a unit circle, is shown in Fig. 6.4b. In this case the strain matrix was:

$$\begin{bmatrix} 2.5 & 0.5 \\ 1 & 3.5 \end{bmatrix}$$

This is an example of general homogeneous rotational strain and as can be seen from Fig. 6.4b, the eigenvectors are not at right angles, nor are they coincident with the axes of principal strain. Measurement of the parallelogram formed by deformation of the unit square shows a factor of magnification of 8.25, the determinant of the matrix. The numerical results, confirming these conclusions are summarized in Table 6.6.

As noted earlier, as with other matrix applications the concepts discussed here can be translated into three dimensions. In this situation the strain ellipse becomes an ellipsoid with three mutually perpendicular axes. As demonstrated in Example 5.11 the strain matrix will now have nine elements, and by definition there will be three associated eigenvalues and corresponding eigenvectors.

Much of the material presented in this chapter and in Chapter 5, section 5.2, is dealt with by Ramsay and Huber (Vol.1, 1983) in a geological context, emphasizing its implications as well as its application to real geological problems. Students whose interests lie in this area of geology, are strongly recommended to study the material presented by Ramsay and Huber, in conjunction with that presented here.

Table 6.6 Calculated values for the principal extensions, slopes, rotations, eigenvalues and corresponding slopes of the eigenvectors for the given matrix. All values correct to 4 dp.

Matrix:			
$\begin{bmatrix} 2.5 & 0.5 \\ 1 & 3.5 \end{bmatrix}$	$(1 + e_1) =$ 3.9118	$\lambda_1 =$ 3.8660	
	$(1 + e_2) =$ 2.1090	$\lambda_2 =$ 2.1340	
	$\tan \theta' = -0.4828$	slope $\lambda_1 =$ 2.7321	
	slope$= 64°\ 44'$	slope $\lambda_2 = -0.7321$	
	rotation$= +4°\ 46'$	not mutually orthogonal	

Table 6.7 Summary of the types of General Homogeneous Strain and their matrices and determinants.

	Plane strain (No area or vol. change)		Non-plane strain (Area or vol. changes)	
	Matrix	Special case	Matrix	
Rotational	$\begin{bmatrix} a & b \\ c & d \end{bmatrix}$	Det = 1 $\begin{bmatrix} 1 & \gamma \\ 0 & 1 \end{bmatrix}$ (Simple Shear)	$\begin{bmatrix} a & b \\ c & d \end{bmatrix}$	0<Det<1 or Det>1
Irrotational	$\begin{bmatrix} a & b=c \\ c=b & d \end{bmatrix}$	Det = 1 $\begin{bmatrix} k & 0 \\ 0 & \end{bmatrix}$ (Pure Shear)	$\begin{bmatrix} a & b=c \\ c=b & d \end{bmatrix}$	0<Det<1 or Det>1

Finally the relationships between the various types of strain discussed and the matrices which describe them, can be summarized in Table 6.7.

6.3.1 Student examples

Q6.8   The position vectors of some points on the circumference of a unit circle, centre (0, 0) are:

$$\begin{bmatrix} x \\ y \end{bmatrix} = \begin{bmatrix} 0.8 \\ -0.6 \end{bmatrix}; \begin{bmatrix} 0 \\ 1 \end{bmatrix}; \begin{bmatrix} 0.2 \\ 0.98 \end{bmatrix}; \begin{bmatrix} 0.4 \\ 0.917 \end{bmatrix}; \begin{bmatrix} 0.6 \\ 0.8 \end{bmatrix}; \begin{bmatrix} 0.8 \\ 0.6 \end{bmatrix}; \begin{bmatrix} 1 \\ 0 \end{bmatrix}$$

Using the strain matrices:

$$\begin{bmatrix} 2 & 2 \\ 2 & 4 \end{bmatrix} \qquad \begin{bmatrix} 1.2 & 0.2 \\ 0.4 & 0.8 \end{bmatrix}$$

(a) find their transformed positions,
(b) calculate the eigenvalues and eigenvectors of the two matrices; and
(c) calculate the length and direction of the principal semi-axis of the strain ellipse for the second matrix as well as its slope before deformation.

Plot this information onto two graphs, one for each matrix, plotting in direction arrows from the origin to each original point, and from each point to its transformed position. Draw in axes representing the values found for parts (b) and (c).

Q6.9   Using the matrix:

$$\begin{bmatrix} 4 & 2 \\ 6 & 5 \end{bmatrix}$$

and coordinates representing a circle of unit radius and centre $(0, 0)$, plot the transformed figures. Find the lengths and orientation of the semi-axes of the strain ellipse. Using the coordinates of the end of one of the semi-axes and the inverse of the matrix, plot the original radius. What is the angle of rotation? Check the graphical solution by calculating the values you have been asked to find.

Q6.10 Using the matrices:

$$\begin{bmatrix} 1.2 & 0.2 \\ 0.4 & 0.8 \end{bmatrix} \quad \begin{bmatrix} 1 & 0.6 \\ 0.6 & 1.5 \end{bmatrix}$$

and the equations given earlier to calculate the lengths and directions of the principal axes of the strain ellipse, demonstrate the differences between homogeneous rotational strain and homogeneous irrotational strain. Calculation to 2 dp should be sufficient for this example.

Q6.11 A matrix representing general homogeneous rotational strain is:

$$\begin{bmatrix} 2 & -2.1 \\ -2 & 4 \end{bmatrix}$$

(a) Express the difference between the length of the semi-axis of the strain ellipse and the eigenvalues as a percentage.
(b) What is the difference between the slope of the eigenvector of the largest eigenvalue and the direction of the major axis of the strain ellipse?
(c) What is the sense of the rotational component and what is its angle?

7

Multivariate statistics

It would not be appropriate or possible to cover all of the applications of linear algebra in multivariate statistics in a text-book such as this. However the widespread use of linear algebra in multivariate statistics and the growing number of applications of these methods in geology for both research and industrial purposes, make some mention desirable. Also it is hoped that students will benefit by seeing the mathematical basis of some methods which are commonly used. In order to address these matters three multivariate statistical methods will be looked at. Those chosen rely heavily on one or more of the linear algebra topics discussed in the earlier chapters, and all are inter-related.

It is assumed that students will already have some knowledge of basic univariate and bivariate statistical methods, their application and their interpretation. For those who do not have the necessary background, or who require some revision, Maroney's book *Facts From Figures* (Maroney, 1953), is an inexpensive but comprehensive introduction. Although not a geological work, it covers all that is needed in a simple, readable and readily understandable fashion. Also it is easily obtained. In particular Chapter 9 which deals with variation and co-variation will repay study, before starting this chapter.

7.1 Some basic matrices

Fundamental to the methods to be considered (as well as many others), are three matrices which are frequently used by statisticians to summarize multivariate data sets:

1. Sums of squares and sums of cross products matrix, frequently designated [S].
2. The dispersion matrix [D], which contains the sums of squares and sums of cross products of the deviations from the mean. Division through by the number of individuals in the sample gives the variance/covariance matrix [V].
3. The correlation matrix [R]. This is a matrix of correlation coefficients, showing the correlation between the variables.

These matrices are square and symmetrical. Additionally the correlation matrix has unity as each element of the principal diagonal, representing self correlations.

Before considering the details of the methods of calculation and demonstrating some applications, it will be useful to define some of the terms which will be used.

- **Population** – all possible individuals or objects having similar attributes and which can be classified together.
- **Sample** – a collection of objects or individuals taken to represent the population.
- **Object or individual** – a single member of the population.
- **Characteristic or variable** – one or more features which can be measured or otherwise quantified for each individual member of the population. In practice these are measured on the individuals which form the sample.

Let us suppose that we have a population from which we have taken a sample, made up of a number of individual objects:

$$O_1 , O_2 , O_3 , \ldots , O_n$$

for each object we have measured several characteristics:

$$v_1 , v_2 , v_3 , \ldots , v_m$$

And that the distribution of each measured characteristic for the population is normal.

This information can be written in matrix form:

	v_1	v_2	v_3	$\ldots$	v_m
O_1	$a_{1,1}$	$a_{1,2}$	$a_{1,3}$	$\ldots$	$a_{1,m}$
O_2	$a_{2,1}$	$a_{2,2}$	$a_{2,3}$	$\ldots$	$a_{2,m}$
O_3	$a_{3,1}$	$a_{3,2}$	$a_{3,3}$	$\ldots$	$a_{3,m}$
.	.	.	.		.
.	.	.	.		.
.	.	.	.		.
O_n	$a_{n,1}$	$a_{n,2}$	$a_{n,3}$	$\ldots$	$a_{n,m}$

$= [\mathbf{A}]$

Note – it is important to remember in most of the applications which follow, the row labels are the object identifications and the column labels are the variable names.

If we perform the matrix operation:

$$[\mathbf{A}]^{\mathrm{T}} . [\mathbf{A}] = [\mathbf{S}]$$

then from the rules of matrix algebra it should be clear that:

1. The diagonal elements $s_{i,j}$ ($i = j$) will contain the sums of squares of the variables

2. The off-diagonal elements $s_{i,j}$ $(i \neq j)$ will contain the sums of cross products of the variables
3. The matrix [S] will be square and of size m, where m is the number of variables measured

The matrix [S] is the sums of squares and sums of cross products matrix.

A mean value (denoted by the symbol $\bar{a}_{v_j}$) for each variable v_1 - v_m can be found by finding the sum of each column of the matrix [A] and dividing by n, the number of objects in the sample, as below:

	v_1	v_2	v_3	$\cdots$	v_m	
o_1	$a_{1,1}$	$a_{1,2}$	$a_{1,3}$	$\cdots$	$a_{1,m}$	
o_2	$a_{2,1}$	$a_{2,2}$	$a_{2,3}$	$\cdots$	$a_{2,m}$	
o_3	$a_{3,1}$	$a_{3,2}$	$a_{3,3}$	$\cdots$	$a_{3,m}$	$= [A]$
.	.	.	.		.	
.	.	.	.		.	
.	.	.	.		.	
o_n	$a_{n,1}$	$a_{n,2}$	$a_{n,3}$	$\cdots$	$a_{n,m}$	

SUM $(s_{v_j(j=1,\,\ldots,\,m)})$ S_1 S_2 S_3 $\cdots$ S_m

MEAN $(\bar{a}_{v_j(j=1,\,\ldots,\,m)})$ $\dfrac{S_1}{n}$ $\dfrac{S_2}{n}$ $\dfrac{S_3}{n}$ $\cdots$ $\dfrac{S_m}{n}$

The mean value for each variable can now be subtracted from the corresponding elements in the matrix [A], to give a new matrix [B]:

$$[B] = \begin{bmatrix} a_{1,1} - \bar{a}_{v_1} & a_{1,2} - \bar{a}_{v_2} & a_{1,3} - \bar{a}_{v_3} & \cdots & a_{1,m} - \bar{a}_{v_m} \\ a_{2,1} - \bar{a}_{v_1} & a_{2,2} - \bar{a}_{v_2} & a_{2,3} - \bar{a}_{v_3} & \cdots & a_{2,m} - \bar{a}_{v_m} \\ a_{3,1} - \bar{a}_{v_1} & a_{3,2} - \bar{a}_{v_2} & a_{3,3} - \bar{a}_{v_3} & \cdots & a_{3,m} - \bar{a}_{v_m} \\ \vdots & \vdots & \vdots & & \vdots \\ a_{n,1} - \bar{a}_{v_1} & a_{n,2} - \bar{a}_{v_2} & a_{n,3} - \bar{a}_{v_3} & \cdots & a_{n,m} - \bar{a}_{v_m} \end{bmatrix}$$

i.e. The matrix [B] will contain the deviations from the mean for every object, for each variable. Then if we multiply:

$$[B]^T \cdot [B] = [D]$$

as before we have:

1. The diagonal elements $d_{i,j}$ $(i = j)$ will contain the sums of squares of the deviations from the mean of each variable.
2. The off-diagonal elements $d_{i,j}$ $(i \neq j)$ will contain the sums of cross products of the deviations from the mean of each variable.
3. [D] will be square and symmetrical and of size m, where m is the number of measured variables.

The matrix [D] is the dispersion matrix, or sums of squares and sums of cross-products of the deviations from the means matrix.

Since the variance is calculated as the sum of squares of the deviations from the mean divided by the number of objects which went to make up the sample and the covariance is the sum of products of the deviations from the mean divided by the number of objects, then the variance/covariance matrix can be obtained by dividing each element of the matrix [D] by n, the number of objects. That is :

$$[D] * \frac{1}{n} = [V]$$

The matrix [V] being the variance/covariance matrix, with the variances in the principal diagonal and the covariances in the off-diagonal elements. The standard deviation (denoted by the symbol σ_{v_j}) for each measured variable:

$$\sigma_{v_1} , \sigma_{v_2} , \sigma_{v_3} , \ldots , \sigma_{v_m}$$

can be calculated by finding the square root of each diagonal element of [V].

Alternatively, they can be calculated from the dispersion matrix [D], when the standard deviations will be given by:

$$\sqrt{\frac{d_{i, j(i=j)}}{n}} = \sigma_{v_j (j=1, \ldots, m)}$$

That is, since the diagonal elements of [D] are the sums of squares of the deviations from the means, then the standard deviation for each variable will be given by the square root of the mean of each element.

Finally, it is convenient to calculate a matrix of normalized variates, i.e. the data is recalculated so that each variable will have a zero mean and a standard deviation of unity. This matrix [C], is calculated by dividing the elements of each row of [B] by the appropriate standard deviation, as below:

$$[C] = \begin{bmatrix} \dfrac{b_{1,1}}{\sigma_{v_1}} & \dfrac{b_{1,2}}{\sigma_{v_2}} & \dfrac{b_{1,3}}{\sigma_{v_3}} & \cdots & \dfrac{b_{1,m}}{\sigma_{v_m}} \\[2ex] \dfrac{b_{2,1}}{\sigma_{v_1}} & \dfrac{b_{2,2}}{\sigma_{v_2}} & \dfrac{b_{2,3}}{\sigma_{v_3}} & \cdots & \dfrac{b_{2,m}}{\sigma_{v_m}} \\[2ex] \dfrac{b_{3,1}}{\sigma_{v_1}} & \dfrac{b_{3,2}}{\sigma_{v_2}} & \dfrac{b_{3,3}}{\sigma_{v_3}} & \cdots & \dfrac{b_{3,m}}{\sigma_{v_m}} \\[2ex] \cdot & \cdot & \cdot & & \cdot \\ \cdot & \cdot & \cdot & & \cdot \\ \dfrac{b_{n,1}}{\sigma_{v_1}} & \dfrac{b_{n,2}}{\sigma_{v_2}} & \dfrac{b_{n,3}}{\sigma_{v_3}} & \cdots & \dfrac{b_{n,m}}{\sigma_{v_m}} \end{bmatrix}$$

If the multiplication operation $[C]^T.[C]$, is now carried out and if n is the number of objects in the original sample, then:

$$[C]^T . [C] = [R] . n$$

where [R] is the correlation matrix. Alternatively, [R] is given by:

$$([C]^T \cdot [C]) * \frac{1}{n} = [R]$$

as before we have:

1. The diagonal elements $r_{i,j}$ ($i = j$) will all be equal to unity, representing self-correlations.
2. The off-diagonal elements $r_{i,j}$ ($i \neq j$) contain the values of the correlation coefficients.
3. The matrix will be square and symmetrical and of size m, where m is the number of variables.

The matrix $[R]$ is the correlation matrix, where the values $r_{i,j}$ are Pearson's product moment correlation coefficient.

Clearly as the number of objects making up the sample of the population and/or the number of variables or characteristics measured increases, hand calculation becomes unattractive and the manipulations are best done by computer. Calculations in many of the examples which follow have been performed by computer using programmes written in Microsoft Quick Basic, and run on an IBM PS/1 computer. Student examples given are capable of solution with patience, using a hand held calculator, although they may be more time consuming than examples given in the earlier chapters.

Example 7.1

Find the sums of squares and sums of cross products matrix for the data given in Student Example Q1.4. In this the data related to the amounts of Pb and Zn measured in a number of sediment samples. They are: (42.7, 14.9), (85.8, 23.2), (57.0, 23.3), (43.0, 16.6), (10.0, 8.0), (11.0, 8.0). Values given in ppm/g of sediment.

Step 1 Write the data in matrix form, with Pb as variable 1 and Zn as variable 2:

Sample No.	Pb	Zn
1	42.7	14.9
2	85.8	23.2
3	57.0	23.3
4	43.0	16.6
5	10.0	8.0
6	11.0	8.0

$= [A]$

Step 2 Write the matrix $[A]$ as its transpose:

Sample No.	1	2	3	4	5	6
ppm/g Pb	42.7	85.8	57.0	43.0	10.0	11.0
ppm/g Zn	14.9	23.2	23.3	16.6	8.0	8.0

$= [A]^T$

Step 3 Pre-multiply the matrix by its transpose:

$$[A]^T . [A] = \begin{array}{cc} \text{Pb} & \text{Zn} \\ \begin{bmatrix} 14503.931 & 4836.69 \\ 4836.69 & 1706.70 \end{bmatrix} & \begin{array}{c} \text{Pb} \\ \text{Zn} \end{array} \end{array}$$

$$= [S]$$

as noted earlier the matrix [S], represents:

$$[S] = \begin{bmatrix} \Sigma x_1^2 & \Sigma x_1 x_2 \\ \Sigma x_1 x_2 & \Sigma x_2^2 \end{bmatrix}$$

The original problem posed in Student Example Q1.4, was to find the constants a and c for the equation of the straight line:

$$x_2 = ax_1 + c$$

where x_2 = Pb and x_1 = Zn (all measurements in ppm/g of sediment). Note: in this and subsequent examples in multivariate statistics we shall revert to using the $x_1 x_2 \ldots x_n$ – plane).

As demonstrated in Chapter 1, using the least squares method, the constants a and c can be found by solving the SLE set:

$$\Sigma x_1^2 a + \Sigma x_1 c = \Sigma x_1 x_2$$
$$\Sigma x_1 a + nc = \Sigma x_2$$

where n is the number of pairs of measurements. Using the information contained in the matrix [S] and finding the column totals for the matrix [A], i.e. Σx_1 and Σx_2, the equations are:

$$14503.931a + 249.5c = 4836.69$$
$$249.5a + 6c = 94.0$$

The solution of which gives:

$$a = 0.191$$
$$c = 8.313$$

hence the equation which describes the observed relationship in the data set is:

$$\text{Pb} = 0.191\text{Zn} + 8.313$$

Example 7.2
The following data relates to the amounts of light hydrocarbon gases methane, ethane and propane (measured in vppm/cc mineral), extracted from fluid inclusions in the mineral fluorite. The samples were collected from a number of localities throughout the Northern Pennine orefield.

	Methane	Ethane	Propane
1	341.0	5.1	2.2
2	907.0	8.3	2.2
3	1009.0	7.3	0.7
4	867.0	6.2	0.7
5	939.0	5.8	0.6
6	508.0	6.2	2.3
7	751.0	5.8	1.4
8	908.0	6.1	1.3
9	889.0	8.5	1.8
10	920.0	6.6	0.7
11	751.0	7.1	1.3
12	804.0	6.0	0.9
13	748.0	6.1	1.4
14	1715.0	10.0	3.3
15	1159.0	8.5	1.7
Means	881.067	6.907	1.50

Calculate the dispersion and correlation matrices for this data set.

The results are as follows (most of the details of the calculation omitted):

Step 1 Calculate the matrix [**B**] (the deviations from the mean):

$$[\mathbf{B}] = \begin{bmatrix} -540.067 & -1.807 & 0.7 \\ 25.933 & 1.393 & 0.7 \\ 127.933 & 0.393 & -0.8 \\ -14.067 & -0.707 & -0.8 \\ 57.933 & -1.107 & -0.9 \\ -373.067 & -0.707 & 0.8 \\ -130.067 & -1.107 & -0.1 \\ 26.933 & -0.807 & -0.2 \\ 7.933 & 1.593 & 0.3 \\ 38.933 & -0.307 & -0.8 \\ -130.067 & 0.193 & -0.2 \\ -77.067 & -0.907 & -0.6 \\ -133.067 & -0.807 & -0.1 \\ 833.933 & 3.093 & 1.8 \\ 277.933 & 1.593 & 0.2 \end{bmatrix}$$

Step 2 Calculate the dispersion matrix [**D**] (using $[\mathbf{B}]^{\mathrm{T}} \cdot [\mathbf{B}] = [\mathbf{D}]$):

$$[\mathbf{D}] = \begin{bmatrix} 1283920.9 & 4569.093 & 819.50 \\ 4569.093 & 25.709 & 7.86 \\ 819.5 & 7.86 & 8.18 \end{bmatrix}$$

Step 3   Calculate the standard deviation for each of the 3 gases:

$$ov_1 \text{ (methane)} = \sqrt{\frac{d_{1,1}}{n}} = \sqrt{\frac{1283920.9}{15}} = 292.566$$

$$ov_2 \text{ (ethane)} = \sqrt{\frac{d_{2,2}}{n}} = \sqrt{\frac{25.709}{15}} = 1.309$$

$$ov_3 \text{ (propane)} = \sqrt{\frac{d_{3,3}}{n}} = \sqrt{\frac{8.180}{15}} = 0.738$$

Step 4   Calculate the matrix [C] (normalized variates matrix), by dividing the elements of [B] by the standard deviations as appropriate:

$$[\mathbf{C}] = \begin{bmatrix}
-1.846 & -1.380 & 0.948 \\
0.089 & 1.064 & 0.948 \\
0.437 & 0.300 & -1.083 \\
-0.048 & -0.540 & -1.083 \\
0.198 & -0.845 & -1.219 \\
-1.275 & -0.540 & 1.083 \\
-0.445 & -0.845 & -0.135 \\
0.92 & -0.616 & -0.271 \\
0.027 & 1.217 & 0.406 \\
0.133 & -0.234 & -1.083 \\
-0.445 & 0.148 & -0.271 \\
-0.263 & -0.693 & -0.812 \\
-0.455 & -0.616 & -0.135 \\
2.850 & 2.363 & 2.437 \\
0.950 & 1.217 & 0.271
\end{bmatrix}$$

Step 5   Calculate the correlation matrix (using $[\mathbf{C}]^{\mathrm{T}} \cdot [\mathbf{C}] * \frac{1}{n} = [\mathbf{R}]$):

$$[\mathbf{R}] = \begin{bmatrix}
1.0000 & 0.7952 & 0.2529 \\
0.7952 & 1.0000 & 0.5420 \\
0.2529 & 0.5420 & 1.0000
\end{bmatrix}$$

NOTE: The values in the principal diagonal of $[\mathbf{R}]*n$ should equal n, the number of samples. If the calculations are performed to a limited number of decimal places, error can accumulate during calculation and the diagonal elements may not be as exact. However the error should not be excessive, and rounding to one less decimal place than that to which the rest of the calculation has been performed, should result in the correct value.

It is convenient at this point to look at one final data set. This again relates to gas released from fluorite samples from the Northern Pennine orefield. It is however a separate data set to that used in Example 7.2, having been collected at a different time and from different locations.

Example 7.3

A second group of samples of fluorite have been analysed for their contained light hydrocarbon gases. The data, in vppm/cc of mineral are:

Mineral No.	1	2	3	4	5	6	7	8	9	10	
Methane	314	333	305	1471	702	1002	824	738	349	406	
Ethane	3.9	1.2	1.5	12.5	7.2	7.7	12.6	9.5	2.3	3.1	$= [A]$
Propane	0.9	1.2	0.4	2.3	2.7	0.8	3.7	3.8	0.9	0.8	

The mean values are: Methane 644.4
Ethane 6.15
Propane 1.75

Calculate the dispersion matrices $[D]$ and the correlation matrix $[R]$.

It is important to note the data layout in this example. The format is different to that used in previous examples, the columns and rows being interchanged. Since the basic matrix operations involve multiplying the matrix and its transpose, the data can be used in this form provided the multiplication operations are reversed as appropriate. Thus, for example, to calculate the sums of squares and sums of cross products matrix $[S]$, we have:

$$[A] \cdot [A]^T = [S]$$

where the matrix $[A]$ is in the row/column order as given above. In the calculation which follows this alternative strategy will be followed.

Step 1 Subtract the mean $(\bar{a}_v v_{m(m=1, 2, 3)})$, from each value as appropriate to give the matrix $[B]$:

$$[B] = \begin{bmatrix} -330.4 & -311.4 & -339.4 & 826.6 & 57.6 & 357.6 & 179.6 & 93.6 & -295.4 & -238.4 \\ -2.25 & -4.95 & -4.65 & 6.35 & 1.05 & 1.55 & 6.45 & 3.35 & -3.85 & -3.05 \\ -0.85 & -0.55 & -1.35 & 0.55 & 0.95 & -0.95 & 1.95 & 2.05 & -0.85 & -0.95 \end{bmatrix}$$

Step 2 Transpose the matrix to give $[B]^T$:

$$[B]^T = \begin{bmatrix} -330.4 & -2.25 & -0.85 \\ -311.4 & -4.95 & -0.55 \\ -339.4 & -4.65 & -1.35 \\ 826.6 & 6.35 & 0.55 \\ 57.6 & 1.05 & 0.95 \\ 357.6 & 1.55 & -0.95 \\ 179.6 & 6.45 & 1.95 \\ 93.6 & 3.35 & 2.05 \\ -295.4 & -3.85 & -0.85 \\ -238.4 & -3.05 & -0.95 \end{bmatrix}$$

Step 3 Multiply the matrix $[B]$ by its transpose, to give:

$$[\mathbf{B}] \cdot [\mathbf{B}]^{\mathrm{T}} = \begin{bmatrix} 1320902.375 & 13063.101 & 2099.6 \\ 13063.101 & 171.965 & 39.545 \\ 2099.6 & 39.545 & 14.585 \end{bmatrix} = [\mathbf{D}]$$

Step 4 Calculate the standard deviations $(\sigma_{v_{m(m=1,\,2,\,3)}})$ for each of the 3 gases:

$$\sigma_{v_1} \text{ (methane)} = \sqrt{\frac{d_{1,1}}{n}} = \sqrt{\frac{1320902.375}{10}} = 363.442$$

$$\sigma_{v_2} \text{ (ethane)} = \sqrt{\frac{d_{2,2}}{n}} = \sqrt{\frac{171.965}{10}} = 4.147$$

$$\sigma_{v_3} \text{ (propane)} = \sqrt{\frac{d_{3,3}}{n}} = \sqrt{\frac{14.585}{10}} = 1.208$$

Step 4 Divide each element of the matrix [**B**] by the standard deviation $(\sigma_{v_{m(m=1,3)}})$ as appropriate (i.e. elements of row 1 by σ_{v1}, row 2 by σ_{v2}, etc.), to give the matrix [**C**]:

$$[\mathbf{C}] = \begin{bmatrix} -0.909 & -0.857 & -0.934 & 2.274 & 0.158 & 0.984 & 0.494 & 0.258 & -0.813 & -0.656 \\ -0.543 & -1.194 & -1.121 & 1.531 & 0.253 & 0.374 & 1.555 & 0.808 & -0.928 & -0.735 \\ -0.704 & -0.455 & -1.118 & 0.455 & 0.787 & -0.787 & 1.615 & 1.697 & -0.704 & -0.787 \end{bmatrix}$$

Step 5 Multiply $[\mathbf{C}].[\mathbf{C}]^{\mathrm{T}}$ and divide each element by n:

$$[\mathbf{C}] \cdot [\mathbf{C}]^{\mathrm{T}} = \begin{bmatrix} 10.000 & 8.667 & 4.784 \\ 8.667 & 10.000 & 7.896 \\ 4.784 & 7.896 & 10.000 \end{bmatrix} = [\mathbf{R}] * n$$

Whence:

$$[\mathbf{R}] = \begin{bmatrix} 1.0000 & 0.8667 & 0.4784 \\ 0.8667 & 1.0000 & 0.7896 \\ 0.4784 & 0.7896 & 1.0000 \end{bmatrix}$$

To eliminate any doubt about the validity of the approach used i.e. starting with the data written in the transposed format and reversing the order of multiplication, students should work through this example after rewriting the data conventionally.

7.1.1 Student examples

Q7.1 Seven samples of sea water collected at Brighton on the south coast of England have been analysed for the elements Mg, Ca and Sr. The results, in ppm/g are given in Table 7.1.

Calculate the sums of squares, sums of cross products matrix, and the dispersion matrix for this data. Students are advised to scale the data for this and the next question (division through by 100 should produce matrices which have reasonable sized values). Keep the results for both questions as they will be needed later in the chapter.

Table 7.1 Data for Q7.1

Mg	Ca	Sr
1300	420	9.4
1350	410	9.3
1280	420	9.0
1270	420	8.9
1330	430	9.4
1330	440	9.4
1250	420	7.7

Q7.2   After experiments designed to simulate the diagenesis of carbonate ooids, the saline fluid remaining in the experimental apparatus gave the data for the elements Mg, Ca and Sr (values in ppm/g) (Table 7.2). Each sample represents fluid from a different experiment.

Calculate the two matrices as in Q7.1.

Table 7.2 Data for Q7.2

Mg	Ca	Sr
134	1740	146
160	880	125
270	730	93
960	520	21
590	580	38
580	620	45
370	730	55
520	610	54
270	670	103
43	1270	120

7.2 Multivariate regression

In Chapter 1, section 1.1.1, when considering the applications of SLE in geology, the example of finding the constants a and c (slope and intercept) of a linear equation was demonstrated. Using differential calculus, it was shown that the values could be found by solving the equation set:

$$\Sigma x_1^2 c_1 + \Sigma x_1 c_0 = \Sigma x_1 x_2$$
$$\Sigma x_1 c_1 + n c_0 = \Sigma x_2$$

each coefficient (x term) being summed from 1 to n, where n is the number of pairs of measurements x_1 and x_2, and unknowns

c_0 (= c the intercept as defined in Chapter 1)
c_1 (= a the slope as defined in Chapter 1)

Using similar arguments it is possible to show that for three variables, the solution of the linear equation:

$$x_3 = c_1 x_1 + c_2 x_2 + c_0 \tag{1}$$

is given by:

$$
\begin{aligned}
nc_0 + \Sigma x_1 c_1 + \Sigma x_2 c_2 &= \Sigma x_3 \\
\Sigma x_1 c_0 + \Sigma x_1^2 c_1 + \Sigma x_1 x_2 c_2 &= \Sigma x_1 x_3 \\
\Sigma x_2 c + \Sigma x_1 x_2 c_1 + \Sigma x_2^2 c_2 &= \Sigma x_2 x_3
\end{aligned}
\tag{2}
$$

where x_1 , x_2 and x_3 are the three measured variables
the sums of products and cross products of the x–terms are the coefficients of the unknowns c_1 and c_2
and c_0 is the constant.

Using matrix inversion, the solution of the SLE (2), for the constant and the unknowns, is given by:

$$
\begin{bmatrix}
n & \Sigma x_1 & \Sigma x_2 \\
\Sigma x_1 & \Sigma x_1^2 & \Sigma x_1 x_2 \\
\Sigma x_2 & \Sigma x_1 x_2 & \Sigma x_2^2
\end{bmatrix}^{-1}
\cdot
\begin{bmatrix}
\Sigma x_3 \\
\Sigma x_1 x_3 \\
\Sigma x_2 x_3
\end{bmatrix}
=
\begin{bmatrix}
c_0 \\
c_1 \\
c_2
\end{bmatrix}
\tag{3}
$$

In this general example ((1) above) x_3 has been chosen as the dependent variable, and therefore x_1 and x_2 are the independent variables. As noted in Chapter 1, the choice of the dependent variable in a regression equation is usually governed by the problem being tackled. In many cases it is clear which variable falls into which category, thus only one equation will be permissible to describe the relationships between the variables. For example in the geological application of trend surface analysis (1) above, would be used to describe a linear surface. In this case the variables x_1 and x_2 will be geographic coordinates locating the position of the geological feature x_3 being measured. Thus the position can be arbitrarily chosen (giving x_1 and x_2) and once fixed, then the measurement of the parameter of geological interest x_3 will be dependent on that position.

We can add further variables to (1) if required and find the appropriate terms for the coefficients of the SLE set by differentiation as before. A more simple approach is to utilize the pattern of coefficients which can be seen to be emerging from the two and three variable cases already studied. Using this pattern which is summarized in matrix form (based on Davis 1973, p.207ff), it is easy to find the appropriate terms for any number of variables.

Notice that the elements of the matrix (except for the coefficients) are obtained by multiplying the row and column labels and prefixing them with the summation sign. Also, the terms in the main body of the matrix to the left of the column of coefficients, are those of the dispersion matrix [D] (the sums of squares and sums of cross products matrix) demonstrated

	x_0	x_1	x_2	x_3	$\ldots$	x_m		x_{m+1}
x_0	n	Σx_1	Σx_2	Σx_3	$\ldots$	Σx_m	c_0	Σx_{m+1}
x_1	Σx_1	Σx_1^2	$\Sigma x_1 x_2$	$\Sigma x_1 x_3$	$\ldots$	$\Sigma x_1 x_m$	c_1	$\Sigma x_1 x_{m+1}$
x_2	Σx_2	$\Sigma x_2 x_1$	Σx_2^2	$\Sigma x_2 x_3$	$\ldots$	$\Sigma x_2 x_m$	c_2	$\Sigma x_2 x_{m+1}$
x_3	Σx_3	$\Sigma x_3 x_1$	$\Sigma x_3 x_2$	Σx_3^2	$\ldots$	$\Sigma x_3 x_m$	c_3	$\Sigma x_3 x_{m+1}$
.	.	.	.	.	$\ldots$	.	.	.
.	.	.	.	.	$\ldots$	.	.	.
x_m	Σx_m	$\Sigma x_m x_1$	$\Sigma x_m x_2$	$\Sigma x_m x_3$	$\ldots$	Σx_m^2	c_m	$\Sigma x_m x_{m+1}$

earlier in this chapter. Therefore using the matrix [D] and calculating the appropriate terms for the column labelled x_{m+1} i.e.

$$\Sigma x_{m+1}, \Sigma x_1 x_{m+1}, \ldots, \Sigma x_m x_{m+1}$$

where $(m+1)$ is the number of variables
and the $(m+1)^{\text{th}}$ variable is the dependent variable
Then the constant c_0 and the coefficients c_1 to c_m for the regression equation can be calculated using matrix inversion as in (3) above.

Example 7.4

Find the multiple regression equation for the data given in Example 7.3 using propane as the dependent variable. The sums of squares sums of cross products matrix is:

$$
\begin{array}{c}
x_1 \\ x_2 \\ x_3
\end{array}
\begin{bmatrix}
x_1 & x_2 & x_3 \\
5473416 & 52693.699 & 13376.6 \\
52693.699 & 550.19 & 147.17 \\
13376.6 & 147.17 & 45.21
\end{bmatrix}
$$

Where x_1 is methane, x_2 is ethane and x_3 is propane.
The sums for each variable are:

$$\Sigma x_1 = 6444 \quad \Sigma x_2 = 61.5 \quad \Sigma x_3 = 17.5$$

for $n = 10$

putting the appropriate values into (3), gives:

$$
\begin{bmatrix}
10 & 6444 & 61.5 \\
6444 & 5473416 & 52693.7 \\
61.5 & 52693.7 & 550.19
\end{bmatrix}^{-1}
\cdot
\begin{bmatrix}
17.5 \\
13376.6 \\
147.17
\end{bmatrix}
=
\begin{bmatrix}
c_0 \\
c_1 \\
c_2
\end{bmatrix}
$$

where c_0 is the constant term and c_1 and c_2 are the coefficients of the methane and ethane terms respectively.
Matrix inversion gives:

$$
\begin{bmatrix}
0.4155 & -0.0005 & -0.0052 \\
-0.0005 & 0.0000 & -0.0002 \\
0.0052 & -0.0002 & -0.0234
\end{bmatrix}
\cdot
\begin{bmatrix}
17.5 \\
13376.6 \\
147.17
\end{bmatrix}
=
\begin{bmatrix}
0.82354 \\
-0.00275 \\
0.4390
\end{bmatrix}
$$

leading to the regression equation:

Propane = $0.82354 - 0.00275$ Methane + 0.439 Ethane

If we change the dependent variable and use methane for example, then all that is needed is to put the appropriate values from the dispersion matrix and from the vector of sums into the equations, to give:

$$\begin{bmatrix} 10 & 61.5 & 17.5 \\ 61.5 & 550.19 & 147.17 \\ 17.5 & 147.17 & 45.21 \end{bmatrix}^{-1} \cdot \begin{bmatrix} 6444 \\ 52693.7 \\ 13376.6 \end{bmatrix} = \begin{bmatrix} c_0 \\ c_1 \\ c_2 \end{bmatrix}$$

which after calculating as before, gives the regression equation:

Methane = $232.519 + 113.837$ Ethane $- 164.696$ Propane

It would not be appropriate to dwell on the significance of these results from either a geological or a statistical stand-point, at any length. Suffice it to say that we can get a rough idea as to the significance of the equations statistically by checking the degree of correlation between the variables. Referring back to Example 7.3 the correlation coefficients for the gases ranged from 0.8667 to 0.4784. For a sample size $n = 10$, the tabulated value for the correlation coefficient is 0.4973 for the 0.1 level of significance, and we might without carrying out further analysis be justified in saying that the regression equation is meaningful. From a geological stand-point (but based on a much larger sample) the correlation between the gases and the regression equations are significant, being related to the conditions of mineral deposition.

In many situations geologists are required to produce contoured representations (maps) of data, frequently based on either small or large numbers of samples. In both cases objective interpretation can be difficult and results are frequently dependent on the experience of the geologist rather than on the data. To overcome this several mathematical techniques have been proposed. Common amongst these is trend surface analysis, which has appeared in many guises over the period of its use. This is regression analysis, x_1 and x_2 being the independent variables (conventionally eastings and northings respectively), with x_3 being the geological parameter of interest and the dependent variable. Clearly a linear regression equation by definition will enable the description (in mathematical terms) of only the simplest surface, represented by a series of parallel lines. Thus in practice it is frequently found necessary to include higher order terms in order to produce a realistic representation of the phenomena being mapped. For example if a simple anticline or syncline is to be mapped then a quadratic surface at least will be required, allowing the description of either a maximum or a minimum. For more complex patterns, say an anticline and syncline occurring together in the area being mapped, a cubic or higher order equation will be required to describe the surface. Indeed the experience of many workers in this field would suggest that equations including terms up to the sixth power are required. Further information and references relating to this topic are available in books on geostatistics, such as that of Davis (1973).

To illustrate the application of a linear surface we consider the data set out in Table 7.3. They refer to a small area ($1km^2$) in Northamptonshire, England, approximately 2 km to the west of the village of Gretton and north east of the township of Corby. This is in an area where ironstone extraction was widespread until the 1970s, when the use of home-produced ore for the manufacture of iron and steel was stopped. The ironstone, found in Jurassic Bajocian (Aalenian) rocks, is extensively developed in Northamptonshire and is generally known under the formation name of the Northampton Ironstone. It is a mixture of siderite and chamosite oolites and mudstones normally between five and seven metres thick. The quality of the ore in terms of iron content varies from top to bottom through the unit, the best quality ore being found in the top two metres.

Early stages of exploration involve a drilling programme, coring the ironstone throughout its thickness and analysing the rock chemically at regular intervals, so that a profile in terms of the iron, phosphorous and sulphur content is obtained. The quarrying strategy developed from this information involves removing a sufficient thickness from the top of the ironstone which when mixed, would give an ore of suitable quality for smelting. At some pits some preliminary treatment was carried out to improve the quality of the ore, prior to shipping it to the furnaces. The data presented here are selected at random from a larger set. They have been much simplified and give for each sample point (located by its Grid reference), the total thickness of the Northampton Ironstone and the thickness of rock to be removed to give a mixture with a minimum of 25% Fe_2O_3. The Grid references refer to Sheet SP99SW and the thicknesses are given to the nearest 6 inches (original measurements in the data set were in feet and inches and have not been converted to their metric equivalent. Also note that during the original survey, the location of the boreholes was given with reference to the National Grid).

Example 7.5

Using the data given in Table 7.3, calculate the equation for the linear surfaces representing the total thickness of the Northampton Ironstone and the workable thickness. Hence draw the surfaces the equations represent for the kilometre grid square whose SE corner has the reference SP920.930. Since all points fall within the same kilometre grid square, it is convenient to use the abbreviated Grid references, for example the first data point will be (0.04, 5.81). This will prevent the values for the sums

Table 7.3 Data relating to the thickness of the Northampton Ironstone for use in Example 7.5. For convenience during calculation the grid values are measured east and north of the point SP920.930. Thickness is given in feet. Points marked with an asterisk may be 'rogue data', see later in this section.

Eastings	Northings	Total thickness	Workable thickness
SP 920.04	935.81	11.0	1.0*
0.23	3.17	16.0	8.0
0.51	0.83	14.0	7.0
0.98	7.56	14.0	6.0
1.61	6.96	12.0	2.0*
1.70	0.80	14.0	7.0
1.70	2.33	13.0	6.0
1.74	5.52	14.5	7.0
2.04	9.39	17.0	6.0
2.42	4.75	15.0	8.0
2.71	8.28	14.0	6.0
2.78	0.81	14.0	7.0
2.87	6.2	15.0	4.0*
2.91	2.04	15.0	8.0
3.03	3.24	14.0	7.0
3.41	7.45	17.5	8.0
4.01	9.82	16.0	7.0
4.12	3.26	15.0	8.0
4.13	2.04	15.0	8.0
4.26	5.55	15.0	7.0
4.27	0.72	15.0	6.0
4.34	6.71	15.0	5.0
5.24	6.98	12.0	6.0
5.35	3.26	16.0	7.0
5.76	2.29	16.0	6.0
6.51	4.64	15.0	5.0
6.58	8.19	16.0	7.0
6.91	6.97	15.0	6.0
8.58	6.99	12.0	5.0
8.76	9.63	16.0	6.0
Sums 109.5	152.19	439.0	187.0

of squares and sums of cross products becoming too large, and leading to problems of round-off error.

(a) For total thickness the sums of squares sums of products matrix is:

$$\begin{bmatrix} 553.643 & 610.091 & 1623.365 \\ 610.091 & 1006.121 & 2241.645 \\ 1623.365 & 2241.645 & 6489.500 \end{bmatrix}$$

(b) For the workable thickness:

$$\begin{bmatrix} 553.643 & 610.091 & 689.590 \\ 610.091 & 1006.121 & 908.950 \\ 689.590 & 908.950 & 1245.000 \end{bmatrix}$$

The sums are as given in Table 7.3. Since the matrix of coefficients which we need contains terms involving x_1 and x_2 (the grid coordinates) and since both sets of data relating to thickness are measured at the same points, there is only one matrix to be inverted which is:

$$\begin{bmatrix} 30 & 109.5 & 152.19 \\ 109.5 & 553.643 & 610.091 \\ 152.19 & 610.091 & 1006.121 \end{bmatrix}$$

The vectors of values for the right hand side of the equation will differ, since they contain values involving the thickness variables. They are:

$$\mathbf{a} \begin{bmatrix} 439.0 \\ 1623.365 \\ 2241.645 \end{bmatrix} \qquad \mathbf{b} \begin{bmatrix} 187.0 \\ 689.59 \\ 908.95 \end{bmatrix}$$

Matrix inversion and multiplying by the vectors **a** and **b** above, give the two linear equations:

and
$$x_3 = 14.009 + 0.125x_1 + 0.033x_2 \qquad (1.)$$
$$x_3 = 6.809 + 0.115x_1 - 0.197x_2 \qquad (2.)$$

Where x_3 is the predicted thickness at points (x_1, x_2), equation (1.) being that for the total thickness and (2.) that for the workable thickness. By taking the grid nodes:

$$1, 2 \; ; \; 1, 4 \; ; \; \ldots \; ; 1, 8$$

$$\cdot \qquad \cdot \qquad \cdot$$

$$\cdot \qquad \cdot \qquad \cdot$$

$$8, 2 \; ; \; 8, 4 \; ; \; \ldots \; ; 8, 8$$

and inserting them in (1.) and (2.) the predicted values for the thickness x_3 can be calculated. This information is plotted on maps in Fig. 7.1. It is interesting to note that the trend of the total thickness is approximately 165° thinning to the SW, whereas the trend of the workable thickness is approximately 60°, thinning to the NW.

Until now we have not been critical about the matrices used to produce the regression equations. Perhaps at this stage it would be wise to remember and apply some of the criteria given in Chapter 4, relating to

(a)

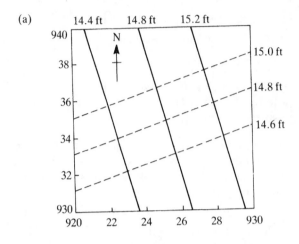

(b)

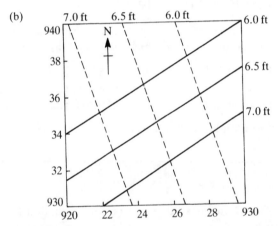

Figure 7.1 Linear trend surfaces of the Northampton Ironstone, data as given in Table 7.3. Grid references refer to 6″ O.S. Sheet SP 99 SW, contour interval in feet.

(a) Solid line – contours based on linear surface for the total data set (Example 7.5). Dashed line – contours based on data set less anomalous points (those indicated with an asterisk in Table 7.3).

(b) Solid line as above, based on thickness of workable Ironstone (mean 25% Fe_2O_3, measured from the top of the unit). Dashed line – same, less anomalous points.

the 'condition' of the equations being used. Some of the criteria given for ill-conditioned matrices are:

1. The absolute value of the determinant is numerically small relative to the coefficients.

2. The elements of the inverse of the matrix are large relative to the coefficients.
3. Small alterations in the coefficients or constants produce large variations in the roots.

Table 7.4 shows some data relevant to these criteria for the last two examples, that is Example 7.4 and Example 7.5. As can be seen from this tabulation, there is every indication that the coefficients of the equations are well conditioned, in spite of the apparently awkward figures in some of the matrices. As a final test the equations for Example 7.5, relating to the Northampton Ironstone were solved using the Gauss–Seidel iterative method when the solutions were produced after between 35 and 45 iterations.

If either of the examples demonstrated had formed part of a proper study, statistical tests would be performed to assess the validity of the conclusions resulting from the data analysis. Although it is not proposed to consider such statistical tests here, it should be noted that when trend surface analysis in particular is used, rogue data points can be problematic. They can upset the interpretation, leading to the suspicion that the surface produced is an artifact. In the example of the Northampton Ironstone demonstrated, there are clearly a small number of data points which may fall into this category. This is particularly so among the data for the workable thickness. Because of this, these data have been re-processed with the rogue points, denoted by an asterisk in Table 7.3, omitted.

The matrices and vectors are given below, those labelled (a) are the originals as calculated and given in Example 7.5, those labelled (b) are those obtained omitting points 1, 5 & 13 from the data set, and recalculating. The two coefficient matrices are:

$$\text{(a)} \begin{bmatrix} 30 & 109.5 & 152.19 \\ 109.5 & 553.643 & 610.091 \\ 152.19 & 610.091 & 1006.121 \end{bmatrix} \quad \text{(b)} \begin{bmatrix} 27 & 104.98 & 133.22 \\ 104.98 & 542.812 & 580.859 \\ 133.22 & 580.859 & 885.484 \end{bmatrix}$$

Table 7.4 Data pertaining to criteria for determining the condition of the coefficient matrices used in Examples 7.4 and 7.5. c_1DV is methane as dependent variable, c_3DV is propane as dependent variable. The final column gives some idea of the variations in the unknowns produced as a result of small alterations in the coefficient matrices.
Other abbreviations – coeff. = coefficient. Tot. T = total thickness.

Example number	Determinant of coeff. matrix	Largest coeff. of matrix	Largest coeff. of inverse
7.4 (c_1DV)	9443	550	0.34
7.4 (c_3DV)	565E+06	547E+04	0.416
7.5 (Tot. T)	991.7E+03	1006.1	0.186

and the four vectors (where, a, indicates total thickness and, b, indicates workable thickness):

(1a) $\begin{bmatrix} 439.0 \\ 1623.365 \\ 2241.645 \end{bmatrix}$ (1b) $\begin{bmatrix} 401.0 \\ 1560.55 \\ 2001.215 \end{bmatrix}$ (2a) $\begin{bmatrix} 187.0 \\ 689.59 \\ 908.95 \end{bmatrix}$ (2b) $\begin{bmatrix} 180.0 \\ 674.850 \\ 864.42 \end{bmatrix}$

Matrix inversion and multiplying by the vectors above as appropriate, leads to the following four linear equations:

with (1a): $x_3 = 14.009 + 0.125x_1 + 0.033x_2$
with (1b): $x_3 = 14.466 - 0.041x_1 + 0.111x_2$
with (2a): $x_3 = 6.809 + 0.115x_1 - 0.197x_2$
with (2b): $x_3 = 7.578 - 0.157x_1 - 0.061x_2$

As before, using the grid node co-ordinate values, the predicted thicknesses have been calculated. These data have been plotted onto maps in Fig. 7.2. It is clear that the omission of the three data points has had a profound effect on the resulting linear surfaces and the isopachs are now almost at right angles to those originally illustrated (Fig. 7.1). This is a good example of the difficulties which can arise when dealing with some apparently innocuous data sets. In this case, although the problem has been shown to lie with the anomalous data points, the lack of data control in the eastern part of the area and in particular the SE corner probably also has some effect. The original data distribution and a possible interpretation of the raw data are also shown on Fig. 7.2, inspection of which indicates quite clearly that a higher order surface will be required to adequately represent the data.

7.2.1 Student examples

Q7.3   A sample of 50 specimens of the Jurassic brachiopod *Kallirhynchia sharpi* (Muir–Wood) collected from the Bathonian, Great Oolite, at Finedon, Northamptonshire were measured. The measurements taken were length, width and thickness and to avoid unnecessarily large values in the matrices, were expressed in cms. If the sums of squares sums of cross products matrix is:

$$\begin{bmatrix} 68.903 & 76.272 & 44.552 \\ 76.272 & 84.738 & 49.537 \\ 44.552 & 49.537 & 30.019 \end{bmatrix}$$

and the sums for each variable are:

length = 56.910 ; width = 62.650 ; thickness = 35.810

Calculate the regression equation relating the three variables, using thickness as the independent variable.

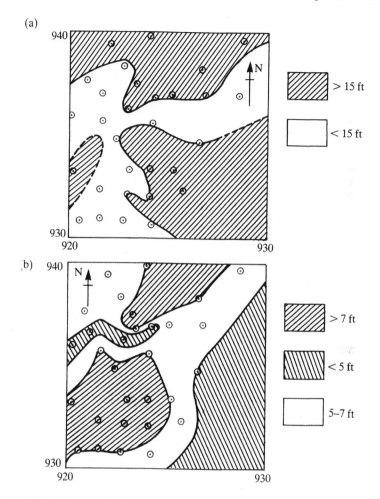

Figure 7.2 Sketch maps showing hand-contoured interpretation of the data in Table 7.3. Location of individual sample points marked.
(a) Total thickness of the Northampton Ironstone
(b) Workable thickness of Northampton Ironstone (mean 25% Fe_2O_3, measured from the top of the unit).

Q7.4 Measurement made of the thickness of the *Productus latissimus* Band, in the Lower Namurian Great Limestone in the area of Weardale, Co. Durham gave the following sums of squares sums of cross products matrix:

$$\begin{bmatrix} 4092.870 & 1136.100 & 189.573 \\ 1136.100 & 505.570 & 53.502 \\ 189.573 & 53.502 & 10.029 \end{bmatrix}$$

And the sums for each variable are:

Eastings = 278.9 Northings = 100.5 Thickness = 13.31

If the thickness was measured in metres, calculate the linear trend equation for the surface. Using the map given as Fig. 7.3 which shows the contoured surface for the original data set and the simplified grid values given along the southern and eastern edges of the map (i.e. those used in the derivation of the matrix and vector above), calculate the expected values and hence draw the linear surface on the map.

7.3 Discriminant functions

A common problem in many areas of data analysis is when there are two or more groups of objects forming samples of two or more populations, for which a number of features have been measured, and the requirement is to classify unknown similar objects with one or other of the populations, based on the same measured features. A multivariate statistical technique which has had extensive application in geology for this purpose, is discriminant analysis. For example in a project in which the author was involved, concerned with demonstrating trade and transportation of Neolithic flint axe-heads, several flint mines and their associated axe-

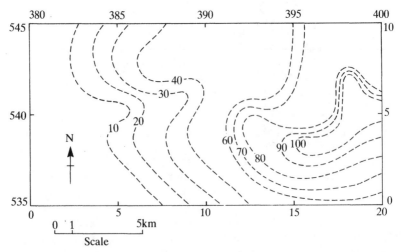

Figure 7.3 Base map for Student example Q7.4, showing contours of the thickness (measured in cm) of the *Productus latissimus* Band in the Lower Namurian, Great Limestone covering an area in Weardale, Co. Durham. Based on unpublished data from the author's field work.
National Grid References are given along the western and northern margins of the map. Simplified grid values along the southern and eastern margins refer to the calculation. Area of map 20 km × 10 km.

factory sites in southern England and in continental Europe were identified. Using sets of flint samples collected from the mine sites, it was shown on the basis of chemical analysis of trace elements in the flint, that they formed distinct populations (Sieveking *et al.*, 1970). Later work applied discriminant function analysis to the data in order to classify archaeological examples of flint axe-heads with their most likely site of origin (Sieveking *et al.*, 1972).

Discriminant function analysis in its simplest form is closely related to regression, since the objective is to calculate a linear function which best separates the groups of objects on the basis of a number of variables measured for all of the individuals of each group. Having calculated the function, unknowns can be assigned to one or other of the groups on the basis of the same criteria. More precisely the discriminant function defines a straight line where clusters of points in multivariate space (the data), have the greatest separation while at the same time have the least inflation, each cluster representing a different group of objects. This is best appreciated by reference to Fig. 7.4, which illustrates the situation where there are two groups and two variables. Note that when plotted, the

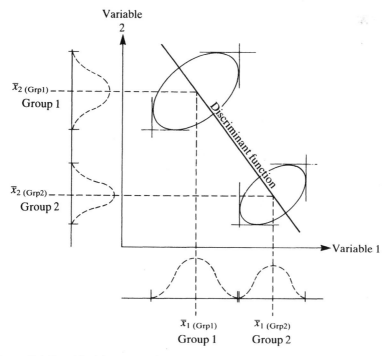

Figure 7.4 Graphical interpretation of the discriminant function in the case of 2 groups with 2 variables. After Harbaugh and Merriam, 1968, Fig. 7–34.

distributions form ellipses. In the case of two groups and two measured variables, the function is:

$$R = X_1 P_1 + X_2 P_2 \tag{1}$$

where X_1 and X_2 are given by:

$$aX_1 + bX_2 = r \atop cX_1 + dX_2 = s \tag{2}$$

the matrix of coefficients is the pooled dispersion matrix for the groups, and the vector for the right hand side of the equations is the differences of the means of the two groups for each variable. The pooled dispersion matrix is calculated as follows:

$$[\mathbf{D}]_{\text{pooled}} = ([\mathbf{D}]_{\text{group1}} + [\mathbf{D}]_{\text{group2}}) / ((n-1)_{\text{group1}} + (n-1)_{\text{group2}})$$

where $[\mathbf{D}]_m$ are the dispersion matrices
for $m = 2$ groups, each group containing n samples (it is not necessary to have the same number of samples for each group). The size of the dispersion matrix (square), will be equal to the number of variables measured. In discriminant function analysis, m will always be less than or equal to the number of variables.

The vector (r s) in the SLE set (2) above is the differences in the means for each group for each variable, i.e.

and
$$r = \text{Mean}_{\text{Var1, Grp1}} - \text{Mean}_{\text{Var1, Grp2}}$$
$$s = \text{Mean}_{\text{Var2, Grp1}} - \text{Mean}_{\text{Var2, Grp2}}$$

The constants X_1 and X_2 in (1) are calculated as:

$$\begin{bmatrix} a & b \\ c & d \end{bmatrix}^{-1} \cdot \begin{bmatrix} r \\ s \end{bmatrix} = \begin{bmatrix} X_1 \\ X_2 \end{bmatrix}$$

The point on the discriminant function which is half-way between the two groups R_0, is found by inserting the mean of the group means for each variable into the discriminant function, (1) i.e.

and
$$P_1 = \tfrac{1}{2} (\text{Mean}_{\text{Var1, Grp1}} + \text{Mean}_{\text{Var1, Grp2}})$$
$$P_2 = \tfrac{1}{2} (\text{Mean}_{\text{Var2, Grp1}} + \text{Mean}_{\text{Var2, Grp2}})$$

Finally we can calculate the point on the discriminant function which is the centre of each of the groups. This is done by inserting the group means into (1). Thus we get:

$$R_{\text{Grp1}} = \text{Mean}_{\text{Var1, Grp1}}X_1 + \text{Mean}_{\text{Var2, Grp1}}X_2$$
$$R_{\text{Grp2}} = \text{Mean}_{\text{Var1, Grp2}}X_1 + \text{Mean}_{\text{Var2, Grp2}}X_2$$

Figure 7.5 shows how the values R_0, R_{Grp1} and R_{Grp2}, relate to the discriminant function.

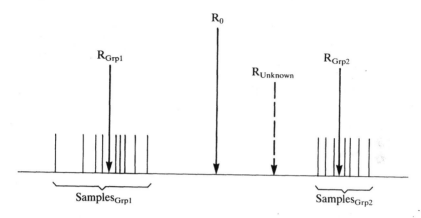

Figure 7.5 Diagrammatic interpretation of R_0, R_{Grp1} and R_{Grp2} etc. After Davis 1973, Fig. 7.5.

All that remains now is to use the discriminant function to classify individual unknown objects to one or other of the groups. This is performed by inserting the values for the variables measured for each unknown object into the discriminant function in turn, and calculating the corresponding value for R. These values can then be compared with the values of R calculated for the centre point of each group and for the point half-way between the groups, to give the necessary classification.

In many computer programmes written to perform discriminant function analysis, R is calculated for each of the original samples. This is used to check the classification by comparing the assignment of individual samples with their original grouping on which the discriminant function was based.

The method and its application can be illustrated by reference to an example.

Example 7.6
Using the data in Table 7.5 calculate the discriminant function for the two Neolithic flint mine sites on the basis of the trace element data given.

Step 1   Using the method outlined earlier in this chapter calculate the dispersion matrices for the two groups:

$$[\mathbf{D}]_{Grp1} = \begin{bmatrix} 108333.695 & 2638.923 \\ 2638.923 & 6303.231 \end{bmatrix} \quad [\mathbf{D}]_{Grp2} = \begin{bmatrix} 17814.223 & 9125.667 \\ 9125.667 & 18720.002 \end{bmatrix}$$

Step 2   Calculate the pooled dispersion matrix:

$$[\mathbf{D}]_{Pooled} = \frac{\{[\mathbf{D}]_{Grp1} + [\mathbf{D}]_{Grp2}\}}{(n_1 + n_2) - 2}$$

Table 7.5 Data relating to Al and Fe (measured in ppm), for samples collected from two Neolithic flint mine sites in southern England. Data taken from Sieveking *et al.*, 1972, Appendix A.

Group 1 (Grimes Graves)		Group 2 (Black Patch)	
Al	Fe	Al	Fe
636	100	325	59
616	105	326	25
605	110	315	72
586	84	422	170
495	80	333	59
668	69	409	67
648	115	316	140
512	53	386	140
635	34	286	72
503	85		
644	100		
863	82		
673	81		
Means 621.846	84.462	346.444	89.333
N 13		9	

where $(n_1 + n_2) - 2 = (13 + 9) - 2 = 20$

which is:

$$[D]_{Pooled} = \begin{bmatrix} 6307.3955 & 588.2295 \\ 588.2295 & 1251.1616 \end{bmatrix}$$

Step 3 Find the differences between the variable means and solve the SLE formed by inserting these values and the pooled dispersion matrix into equation set (2):

$$r = 621.846 - 346.444 \qquad s = 84.462 - 89.333$$
$$= 275.402 \qquad\qquad = -4.871$$

which gives:

$$\begin{bmatrix} 6307.3955 & 588.2295 \\ 588.2295 & 1251.1616 \end{bmatrix}^{-1} \cdot \begin{bmatrix} 275.402 \\ -4.871 \end{bmatrix} = \begin{bmatrix} X_1 \\ X_2 \end{bmatrix}$$

$$= \begin{bmatrix} 0.04605 \\ -0.02554 \end{bmatrix}$$

Hence, (1), the discriminant function is:

$$R = 0.04605P_1 - 0.02554P_2$$

Step 4 Calculate the mid-point for the two variables, hence calculate R_0, R_{Grp1} and R_{Grp2}:

Mid-point for variable 1:

$$\frac{(621.846 + 346.444)}{2} = 484.145$$

Mid-point for variable 2:

$$\frac{(84.462 + 89.333)}{2} = 86.8975$$

Hence:

$$R_0 = 0.04605 * 484.145 - 0.02554 * 86.8975$$
$$= 22.295 - 2.2194$$
$$= 20.0756$$

Putting the means for each variable per group into the discriminant function:

$$R_{Grp1} = 0.04605 * 621.846 - 0.02554 * 84.462$$
$$= 26.479$$
$$R_{Grp2} = 0.04605 * 346.444 - 0.02554 * 89.333$$
$$= 13.672$$

Using the discriminant function and the information just calculated relating to the mid-points of each group and the point mid-way between them, we are now in a position to classify unknown objects. Let us suppose that four axe heads have been analysed to give the following data for Al and Fe: (343, 110); (598, 150); (644, 78) and (464, 105), all values in ppm. To which mine site do they classify?

For axe head 1:

$$R = 0.04605 * 343 - 0.02554 * 110$$
$$= 12.99$$

For axe head 2:

$$R = 0.04605 * 598 - 0.02554 * 150$$
$$= 23.71$$

For axe head 3:

$$R = 0.04605 * 644 - 0.02554 * 78$$
$$= 27.66$$

For axe head 4:

$$R = 0.04605 * 464 - 0.02554 * 105$$
$$= 18.69$$

Since the values for R calculated for axe heads 2 and 3 are greater than the mid-point on the discriminant function and are close to R_{Grp1} in value, then they can be classified with the Grimes Graves source. Similarly axe head

1 classifies with the Black Patch source. Axe head 4 on the other hand although belonging to the second group is very close to the mid-point between the groups, therefore unless further evidence is forthcoming, it might be better to defer classification.

In the situation where there are more variables and groups and provided that there are at least as many variables as groups, then the procedure is exactly the same, the pooled dispersion matrix will have the same dimension as the number of variables and there will be the same number of terms in the discriminant function. For three or more groups, there will be as many pooled dispersion matrices and discriminant functions as possible pair-wise comparisons between the groups.

7.3.1 Student example

Q7.5 Using the dispersion matrices calculated in Q7.1 and Q7.2, calculate the discriminant function, and find R_0, R_{Grp1} and R_{Grp2}, where Group 1 is the sea water and Group 2 is the brines after the experiments. To which group do the data given in Table 7.6 belong?

7.4 Principal components analysis

As stated earlier in this chapter, implicit in the ideas behind multivariate statistical methods, is that the distribution of the population from which the sample was taken is multivariate normal. Thus if two variables which have normal distributions are plotted as in Fig. 7.4, they will define an ellipse representing a bivariate normal population. Contours parallel to the circumference of this ellipse, will join points of equal population density. Similarly three or more variables will plot as ellipsoids representing mutivariate normal populations. Principal components analysis (PCA), is concerned with defining the lengths and directions of the axes of such multivariate normal distributions.

In discriminant function analysis the similarity to regression analysis was noted. In PCA the similarity is again present. That is the axes pass as close as possible to members of the sample which they represent, so that the sums of squares of the deviations from the axes are a minimum. Each

Table 7.6 Data for Q7.5

	Mg	Ca	Sr
1	120	600	90
2	150	400	10
3	1110	600	100
4	250	745	98
5	1300	425	8

other component identified obeys the same criteria, subject to the single restriction that it is at right angles to all other components.

In its simplest form the paradigm of the method of principal components is to find the lengths and directions of the long and short axes of an ellipsoid, where the axes are mutually orthogonal. As we demonstrated in Chapter 6 the ellipsoid defined by a symmetrical matrix will by definition have mutually orthogonal axes whose lengths and directions will be defined by the eigenvalues and eigenvectors of the matrix. These are the **principal components**. It should be noted that in some texts the eigenvectors are known as the component loadings. The matrices used in this analysis are either the dispersion matrix or the correlation matrix. Both matrices are real and symmetrical and by definition have real eigenvalues, factors which help simplify the calculations.

PCA is one of a number of closely related techniques including factor analysis and replaces the much older method of cluster analysis, for the investigation of relationships between variables. Although it was developed as an aid to the interpretation of psychological data, its ability to isolate important factors from multivariate data sets has been exploited by scientists in other disciplines including geology. Since the starting point for many factor-analytical techniques is PCA and since its derivation and interpretation is more straightforward than other methods, it will be considered here.

As in many situations the explanation of the method and the interpretation of the results, is best performed with reference to an example. To this end, the data set which will be examined is set out in Table 7.7, which relates to the minor elements Ca, Na and K represented by the percentage oxide, found during electron microprobe analysis of grains of chlorite in thin sections of sedimentary rock. These data are a sub-sample of a larger unpublished data set from the author's research.

Example 7.7

Perform a principal components analysis of the data given in Table 7.7.

Step 1 Calculate the dispersion matrix for the data as outlined in section 7.1:

$$[\mathbf{D}] = \begin{bmatrix} 7.087 & 1.052 & -0.394 \\ 1.052 & 5.208 & 0.631 \\ -0.394 & 0.631 & 2.214 \end{bmatrix}$$

At this point it is useful to calculate the contribution of each variable to the overall variation in the data set. Remembering that the elements of the principal diagonal of $[\mathbf{D}]$ contain the sums of squares of the deviation from the mean for each variable, and that:

Table 7.7 Data for the trace oxides CaO, Na$_2$O and K$_2$O for 12 samples of chlorite from sedimentary rocks. Values given as a percentage of the mineral.

	CaO	Na$_2$O	K$_2$O
	0.337	1.268	0.166
	0.3	1.03	0.08
	0.768	1.722	0.345
	2.371	1.662	0.352
	2.394	1.519	0.345
	0.173	0.591	0.147
	0.593	2.447	1.37
	0.298	1.417	0.638
	0.204	0.667	1.372
	0.599	0.627	0.15
	0.319	0.993	0.352
	0.173	2.748	0.242
Means	0.711	1.391	0.463

$$\text{Variance} = \frac{\text{Sum of the deviations from the mean}}{\text{number of samples}}$$

then the variation contributed by each oxide, expressed as percentage will be given by:

$$\text{Var}_i = \frac{d_{i, j\,(i\,=\,j)}}{\Sigma d_{i, j\,(i=j=1,\,2,\,3)}} * 100$$

In statistical literature $\Sigma d_{i, j\,(i=j=1,\,...,\,n)}$, is sometimes referred to as: 'the sum of the trace of the matrix'.

$$\Sigma d_{i, j\,(i=j=1,\,...,\,n)} = 7.087 + 5.208 + 2.214$$
$$= 14.509$$

and the values are:

$$\text{Var}_{\text{CaO}} = \frac{7.087}{14.509} * 100$$
$$= 48.85\%$$

$$\text{Var}_{\text{Na}_2\text{O}} = \frac{5.208}{14.509} * 100$$
$$= 35.9\%$$

$$\text{Var}_{\text{K}_2\text{O}} = \frac{2.214}{14.509} * 100$$
$$= 15.25\%$$

Step 2 Find the eigenvalues and eigenvectors for the matrix [**D**]. For this example a computer program written using the Jacobi method for symmetrical matrices was used for ease of computation. Similar results would have been obtained using the technique described in section 6.2.3. The data in Table 7.8 follows the convention adopted in most statistical texts.

Table 7.8 Data for Step 2 of Example 7.7

	Evect 1	Evect 2	Evect 3
Variable 1 (CaO)	0.915	−0.385	0.123
Variable 2 (Na_2O)	0.404	0.885	−0.231
Variable 3 (K_2O)	−0.020	0.261	0.965
Eigenvalue	7.56	4.936	2.013
Variation accounted for	52.11	34.02	13.87

The last row of the tabulated results is simply the eigenvalue expressed as a percentage of the total (should equal the sum of the trace of the matrix from which they were calculated – a good check on the accuracy of the calculation, see section 6.2.1). The percentages indicate the amount of the variation in the original data accounted for by each eigenvalue. In order to use this information, we need to know which of the variables each eigenvalue is associated with. This is achieved by inspecting the eigenvectors, associated with each variable. The largest numerical value for each eigenvector has the same row index as the variable which contributes the amount of variation accounted for. From Table 7.8 we can say that 52.11% of the variation in the data is associated with Variable 1, CaO (Eigenvector 1, value 0.915), 34.02% of the variation is associated with Variable 2, Na_2O (Eigenvector 2, value 0.885) and 13.87% is associated with Variable 3, K_2O (Eigenvector 3, value 0.965). It should be noted that in this example the order of the eigenvalues is the same as that of the variables. This is fortuitous and in other examples the 'largest' eigenvector may be associated with a variable other than the first.

In this example PCA has told us little more than we might have suspected if we had simply considered the contribution of each variable to the overall variation in the data, as displayed by the dispersion matrix. The small differences between the two interpretations which are apparent will be explained later.

The values in the body of the table (eigenvectors), are often referred to as 'factor loadings'. The reason for this follows from the fact that multiplication of the raw data matrix as laid out in Table 7.7 (rows = samples, columns = variables), by the transpose of the matrix of eigenvectors, gives a matrix of loadings (or scores) by projecting the original data onto the three principal component axes. That is:

$$[A]^T . [E]^T = [Y]$$

where $[A]^T$ is the transpose of the data matrix as given in section 7.1,
$[E]^T$ is the transpose of the matrix of eigenvectors as above and
$[Y]$ is the matrix of loadings (scores) on the principal axes

For sample 1:

$$[0.337 \quad 1.268 \quad 0.166] . \begin{bmatrix} 0.915 & 0.404 & -0.02 \\ -0.385 & 0.885 & 0.261 \\ 0.123 & -0.231 & 0.965 \end{bmatrix}$$

$$= [-0.15941 \quad 1.21998 \quad 0.48440]$$

The three values calculated above are the coordinates of the position of Sample 1, relative to the three factor axes. The complete matrix of loadings (or scores) correct to 2 dp is given below. In this matrix, values y_1 are the coordinates for Axis I, y_2 are the coordinates for Axis II and y_3 are the coordinates for Axis III, each row representing a single sample:

$$[Y] = \begin{bmatrix} y_1 & y_2 & y_3 \\ -0.16 & 1.22 & 0.48 \\ -0.11 & 1.01 & 0.34 \\ 0.08 & 1.76 & 0.77 \\ 1.57 & 2.35 & 0.73 \\ 1.65 & 2.23 & 0.68 \\ -0.05 & 0.56 & 0.29 \\ -0.23 & 2.09 & 1.95 \\ -0.19 & 1.23 & 0.98 \\ 0.10 & 0.36 & 1.49 \\ 0.33 & 0.76 & 0.30 \\ -0.05 & 0.93 & 0.59 \\ -0.87 & 2.45 & 0.95 \end{bmatrix}$$

Figure 7.6a shows plots of the projections of this data onto the three axes and in Fig. 7.6b the raw data (Table 7.7) is plotted for comparison. We should also note that since the principal axes are mutually orthogonal, the correlation between the variables will be zero. This factor accounts for the differences between the variances calculated for each variable and the variation accounted for by each eigenvector. Further, the differences between the plots illustrated in Figs 7.6a and 7.6b, where the factor loadings plot is much 'tighter' than that for the raw data, can also be attributed to a redistribution of the variation. To demonstrate that the eigenvectors are orthogonal, the method described in section 5.1.2 can be used. Cross multiplication between pairs of columns of eigenvectors and summing, should give zero values for orthogonal vectors. Using eigenvectors 1 and 2:

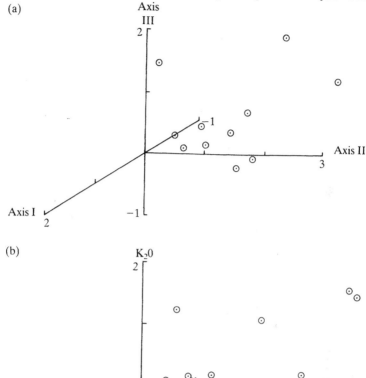

Figure 7.6 (a) The projection of the loadings or scores for the chlorite samples (see Example 7.7), plotted onto the 3 factor axes.
(b) The raw chlorite data (Table 7.7) plotted onto the 3 element axes, for comparison (mean value denoted by ×).

$$(0.915 * -0.385) + (0.404 * 0.885) + (-0.02 * 0.261)$$
$$= -0.3523 + 0.3575 - 0.0052$$
$$= 0.000$$

Students should note that there is little point in performing this calculation to any greater degree of accuracy, this is because the eigenvectors have been calculated using an iterative method, giving results which are probably only accurate to 3 dp, which is however, sufficient for the present purpose.

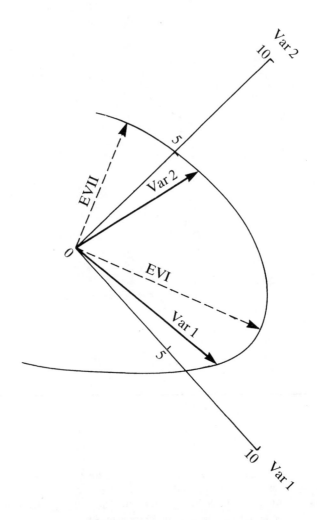

Figure 7.7 Graphical representation of the variances and covariances for variables 1 and 2 (Example 7.7) (CaO and Na$_2$O respectively) and the associated principal components (EV 1 and EV 2), for the chlorite data.

The relationship between the contribution to the total variation in the data for each variable and the eigenvalues and eigenvectors can also be shown graphically. This is demonstrated in Fig. 7.7, by showing pair-wise relationships using the variables CaO and Na$_2$O. On the diagram the x_1–axis represents CaO and the x_2–axis represents Na$_2$O. In this figure, the origin of the four vectors is the point $(0, 0)$, and their final positions were plotted as follows.

The four elements $d_{1,1}$, $d_{1,2}$, $d_{2,1}$ and $d_{2,2}$ from the matrix $[D]$ are used to plot the final positions of the vectors labelled Var 1 and Var 2. The direction of the vector labelled EV 1 was found by plotting the values for the first eigenvector (rounded to 1 dp) for variables 1 and 2 and its length is the numerical value of the first eigenvalue scaled accordingly, i.e. 7.56 units. EV 2 was plotted similarly, using the data relating to the second eigenvector.

As can be seen from Fig. 7.7, the two eigenvectors are at right angles and the four vectors plot on the circumference of an ellipse with centre (0, 0), whose semi-axes are defined by the eigenvalues. In effect the analysis has rotated the variance/covariance plots so that they are at right angles to one another, as expressed by the principal components (eigenvectors).

Further details of the subtleties of PCA and Factor analysis and their application, can be found in most geostatistical texts (e.g. Davis, 1973), or in text-books dealing specifically with multivariate analysis; Moroney's *Facts from Figures*, referred to earlier in this chapter does not cover multivariate procedures. As has been noted in the discussion above, the principal problem which many students encounter is one of nomenclature and the use of symbols. The apparent lack of conformity of the latter can be particularly misleading and off-putting when apparently straightforward operations in linear algebra, become clouded in a mess of conflicting symbols. Any students wishing to explore Factor analysis further are advised to familiarize themselves with the calculation procedures outlined above (using if necessary a good computer programme to calculate the eigenvectors), before using one of the many readily available Factor analysis packages. Thereby the results will not be something produced 'by magic', and there will be a better than average chance of any problems which might arise, being picked up before publication.

7.4.1 Student example

Q7.6 Calculate the dispersion matrix for the following data, and use it as the basis of principal components analysis. Plot the variances and the principal components for the two variables as vectors with origins (0, 0), and show that they all plot on the circumference of an ellipse. The data are:

(17, 16); (23, 10); (14, 11); (15, 19); (9, 52); (10, 56); (26, 14); (10, 18); (11, 20); (21, 14); (9, 24); (11, 12); (11, 17); (17, 14); (26, 14); (12, 88); (11, 27)

Values are in ppm/g sediment and refer to the elements Cu and Zn, detected in shales interbedded with limestone. As before, they are part of a larger data set from the author's research.

8

Review of geological applications

Throughout the text, the examples used to demonstrate the basic techniques of linear algebra, have been as far as possible selected from some relevant geological application. Of necessity the examples have been relatively simple and in some cases trivial. They were chosen first and foremost to show how the calcuation should be performed, their geological relevances being secondary. However it is hoped that the reader has been given an opportunity to see how linear algebra might be applied in modern geology. Clearly there are limitations in the approach adopted here, and there are many instances of the use of linear algebra in the solution of geological problems which have been neglected. In this final chapter a selection of these will be looked at, by reviewing some of the more readily accessible published books and papers.

A good starting point for this review is the work of the American geologists Graeme Bonham-Carter, John Davis, John Harbaugh and Dan Merriam. In three classic text books published between 1968 and 1973, they covered many areas of geology where mathematical applications were then currently in vogue, including several involving the application of linear algebra. Much of what is covered in these books is as relevant today as it was when it was first published.

In the earliest of these books (Harbaugh and Merriam, 1968), topics dealt with included trend surface analysis, discriminant function analysis and factor analysis. The authors also include extensive reviews of much of the early mathematical geology literature, which if followed up, would enable readers to acquire an historic perspective on the development of the methods in geological studies. The same topics were also covered by Davis (1973), where the mathematical basis and methodology is more fully explained with many worthwhile examples as well as computer programmes suitable for their solution. In Chapter 4 (p.127ff), Davis also provides an excellent introduction to matrix algebra itself.

The work by Harbaugh and Bonham–Carter (1970), is primarily concerned with the application of mathematical models, through computer simulation, to geological problems. Several of the areas they deal with involve linear algebraic methods, some of which we will consider in the next two sections.

8.1 Markov processes and the simulation of sedimentary sequences

In the study of sedimentary processes and in particular the simulation of sedimentary sequences, Harbaugh and Bonham-Carter (1970, Ch. 4), discuss ideas relating to the controlling mechanisms in cyclic sedimentation. Important in this context is the notion of processes which are not totally independent, but whose outcome is dependent on previous events. These are termed Markov processes in honour of the Russian mathematician Markov, who was one of the earliest workers to study these interdependent events processes earlier this century. Markov processes are part of probability theory, and are common in some physical systems. They can be classified in a simple fashion by reference to the length of influence of preceding events, thus a one-stage Markov process would describe the situation where only the event immediately preceding had influence. For a more detailed discussion of the basic theory see for example Gray (1967, Ch. 7).

Central to the practical application of the theory is the transition probability matrix which is a convenient form of expression for most applications. This matrix contains values derived from observed data, and as such describes how the system might change at each trial. In other words it contains information relating to the probability of the system moving from one state to another. The probability values are measured in the range 0 to 1, and the row totals of such a matrix will always be unity.

Let us suppose that we have an n–state system n being finite, the states being A, B, ... , N, then the elements p_A, p_B, ... , p_N of any row $(1, ... , n)$, of the matrix will be the probability that the system will move to any other of the possible states (including self transitions). Since the matrix has no negative values and has unit row totals it is called a **stochastic matrix**, also as $i = j = n$, it is a square matrix. The probability of the system being in a particular state after m–transitions will be given by the mth power of the transition matrix. Successive powering of a transition probability matrix will generally lead to a limiting matrix where the elements of each column are equal. This unique row vector represents the proportions of the n–states which would occur after an infinite number of transitions. It is also independent of the starting state. An alternative method for the calculation of this vector is possible using the eigenvalues and eigenvectors of the transition probability matrix. The details of this calculation and a worked example are given by Gray (1967, p.180–2).

In the application of Markov processes to the study of sedimentation, the transition probability matrix is normally compiled from the stratigraphical sequence being studied. In this the first objective is to compile a matrix showing the number of times one lithology passes into

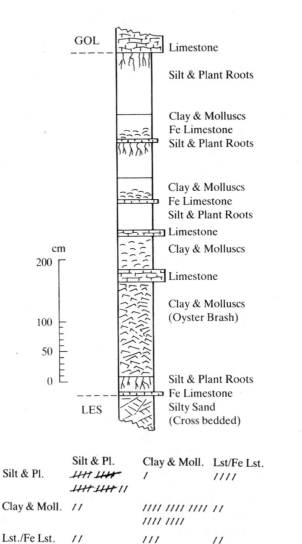

	Silt & Pl.	Clay & Moll.	Lst/Fe Lst.
Silt & Pl.	⊬⊬⊬ ⊬⊬⊬ ⊬⊬⊬ ⊬⊬⊬ //	/	////
Clay & Moll.	//	//// //// //// // //// ////	
Lst./Fe Lst.	//	///	//

	Silt & Pl.	Clay & Moll.	Lst/Fe Lst	Total
Silt & Pl.	22	1	4	27
Clay & Moll.	2	24	2	28
Lst./Fe Lst.	2	3	2	7

Data for 10cm sampling interval

Figure 8.1 A simplified measured stratigraphic sequence in the Upper Estuarine series (Jurassic, Bajocian), at Cranford St.John Ironstone Pit, Cranford St.John, near Kettering, Northamptonshire. Showing the results of a sampling scheme, using a 10 cm interval, for Markov chain analysis.
LES – Lower Estuarine series, GOL – Great Oolite Limestone (Bathonian).

another, including self transitions. This is normally performed at a set interval (measured in terms of a predetermined thickness), although in some studies this is ignored and only bed transitions noted; see Harbaugh and Bonham–Carter (1970, p.105–109) for details. Figure 8.1 illustrates this using a section measured in the Jurassic, Upper Estuarine series, in which there are a number of simple cycles showing marine/estuarine beds passing into non-marine silts which supported plant growth.

Using the data given in Fig. 8.1 for a 10 cm sample interval, the transition probabilities for each row of the data matrix can be calculated by dividing through by the row totals:

	Silt with plants	Clay with molluscs	Limestone/Ironstone
Silt +	$\frac{22}{27}$	$\frac{1}{27}$	$\frac{4}{27}$
Clay +	$\frac{2}{28}$	$\frac{24}{28}$	$\frac{2}{28}$
Lst/Felst	$\frac{2}{7}$	$\frac{3}{7}$	$\frac{2}{7}$

Which gives the transition probability matrix:

$$\begin{bmatrix} 0.81 & 0.04 & 0.15 \\ 0.07 & 0.86 & 0.07 \\ 0.29 & 0.42 & 0.29 \end{bmatrix}$$

To use this information to simulate the sequence, the matrix is converted into the **cumulative transition probability matrix** (CTP matrix) by summing along the rows:

$$\begin{bmatrix} 0.81 & 0.85 & 1.00 \\ 0.07 & 0.93 & 1.00 \\ 0.29 & 0.71 & 1.00 \end{bmatrix}$$

This matrix is then used in conjunction with some means of generating random numbers in the range 0 to 1. Thus by choosing a starting lithology l_s (row label), the random number is used to select the next lithology l_1 (column label). By going to the row indicated and obtaining a second random number a further lithology is obtained:

We choose limestone/ironstone as the starting state, and let us suppose that our first random number is 0.125. Then, using the third row of the CTP matrix and since 0.125 is in the range 0 to 0.29 then the next lithology will be silt with plants. We now consider the first row of the CTP matrix and generate a second random number. Let us suppose this number is 0.78, then, since it falls in the range 0 to 0.81 (first value of first row of matrix), then the next lithology is again silt with plants (can be termed a self transition). Thus we can record the first three lithologies of the simulated sequence:

l_s	Limestone/Ironstone	10 cm (base of sequence)
l_1	Silt with plant rootlets	10 cm
l_2	Silt with plant rootlets	10 cm

Continuing this process will generate as many transitions as required to give a new stratigraphical sequence, which although it will not be exactly the same as that from which the original transition probability matrix was generated, will still exhibit the same transition probabilities. If as in the example used above, the original transition probability matrix was found by using a set sampling interval, then each simulated transition (lithology) will have the same thickness as the sample interval chosen. Where transitions from one lithology to another are used to compile the matrix, no information relating to thickness will be implied. In which case, the thickness of the simulated lithologies will have to be assigned by some other, independent means.

8.2 Differential and partial differential equations

In Chapter 5 of *Computer Simulation in Geology* (p.169ff), Harbaugh and Bonham-Carter (1970) consider the problems associated with the solution of differential and partial differential equations, by considering suitable numerical algorithms for their solution. This is an area of mathematics which is extremely important to physical scientists including geologists. For example any attempt to model flow or transportation, will usually involve the solution of differential equations and in particular, the Laplace equation. The principal method discussed by Harbaugh and Bonham–Carter (1970, p.169ff), is that of finite-differences. The method has three steps:

1. form a finite-difference quotient for the derivative.
2. formulate a set of SLE which will enable the unknowns to be evaluated – the number of equations will be equal to the number of unknowns.
3. solve the SLE set – the method of solution recommended by the authors is the Gauss–Seidel iteration, which was outlined in Chapter 1, section 1.2.3.

The method is best illustrated by reference to a simple example. Consider the linear differential equation:

$$\frac{d^2y}{dx^2} = ax$$

For reference a table of general finite-difference approximations for a number of differentials is given in Table 8.1. Selecting the appropriate approximation, the linear differential given can be translated into a finite-difference quotient:

$$y_{j-1} - 2y_j + y_{j+1} = ax_j(\Delta x)^2$$

Suppose we wish to evaluate x in the range c to d, then the next step will be to set an interval for Δx which will be small but finite, let us call this value s. This will give:

$$\frac{(d - c)}{s} = k$$

Table 8.1 General finite-difference approximations (Central-Difference Formulae of Order $O(h^2)$, for differential equations up to order four. After Harbaugh and Bonham-Carter (1970, p.184).

$$\left(\frac{dy}{dx}\right)_j \simeq \frac{y_{j+1} - y_{j-1}}{2\Delta}$$

$$\left(\frac{d^2y}{dx^2}\right)_j \simeq \frac{y_{j+1} - 2y_j + y_{j-1}}{(\Delta x)^2}$$

$$\left(\frac{d^3y}{dx^3}\right)_j \simeq \frac{y_{j+2} - 2y_{j+1} + 2y_{j-1} - y_{j-2}}{2(\Delta x)^3}$$

$$\left(\frac{d^4y}{dx^4}\right)_j \simeq \frac{y_{j+2} - 4y_{j+1} + 6y_j - 4y_{j-1} - y_{j-2}}{(\Delta x)^4}$$

the values c, d and s should be chosen so that k is an integer value
The values of x will be indexed j ($j=0, \ldots, k$), as follows:

$$x_0 = c, \; x_1 = c + (\Delta x), \; x_2 = c + 2(\Delta x), \ldots, x_k = d$$

or simply
$$x_j = c + j(\Delta x)$$

These values can now be inserted into the finite-difference equation, using the index j from 1 to k (in this example $k = 4$):

$$y_0 - 2y_1 + y_2 = ax_1$$
$$y_1 - 2y_2 + y_3 = ax_2$$
$$y_2 - 2y_3 + y_4 = ax_3$$
$$y_3 - 2y_4 + y_5 = ax_4$$

thus we have a set of four equations with six unknowns. These equations are underdetermined and can only be solved for four unknowns, however if we set the boundary conditions this operation will provide the two values for y necessary to solve the equations. These are y_0 and y_5, which for this illustration will be set to b_1 and b_2 respectively. These are inserted into the equations to give:

$$-2y_1 + y_2 = ax_1 - b_1$$
$$y_1 - 2y_2 + y_3 = ax_2$$
$$y_2 - 2y_3 + y_4 = ax_3$$
$$y_3 - 2y_4 = ax_4 - b_2$$

which can now be solved to give values for y, using a method such as the Gauss–Seidel iteration.

In many practical situations there is more than one independent variable, thus for example as we saw earlier in section 7.2, if our problem was two dimensional with x as the easting and y as the northing then the function representing some measured variable would be:

$$z = f(x, y)$$

If we have an arbitrary grid forming a series of cells, each cell of size Δx (easting) by Δy (northing) with an index i representing increments east, an index j representing increments north, with z being the parameter measured for each cell. The derivatives of z will be partial derivatives with respect to either x or y, which can be written as:

$$\frac{\delta z}{\delta x} \text{ and } \frac{\delta z}{\delta y}$$

Higher-order partial derivatives can be written in the usual way. Using the tabulated finite-difference approximations (Table 8.1), we get:

$$\frac{\delta z}{\delta x} = \frac{z_{i+1, j} - z_{i-1, j}}{2\Delta x}$$

and

$$\frac{\delta z}{\delta y} = \frac{z_{i, j+1} - z_{i, j-1}}{2\Delta y}$$

Also, the second order partial derivatives are:

$$\frac{\delta^2 z}{\delta x^2} = \frac{z_{i+1, j} - 2z_{i, j} + z_{i-1, j}}{\Delta x^2}$$

and

$$\frac{\delta^2 z}{\delta y^2} = \frac{z_{i, j+1} - 2z_{i, j} + z_{i, j-1}}{\Delta y^2}$$

It should be noted that there are two indices in these equations and when y is constant j is constant and conversely, when x is constant i remains constant. These finite-difference approximations are used in a similar fashion to the example given earlier with some important differences.

Students wishing to acquire further insights into the geological applications of partial differentials and the finite-difference method outlined above, are recommended to study Harbaugh and Bonham-Carter (1970, p.184ff). It is also suggested that since their work is geared primarily to geological applications and is by no means exhaustive, other mathematical texts are consulted. Mathews (1992, Ch. 6 and Ch. 9), presents a good summary of the methods which can be used, with many solved examples. Also given are outline computer programmes, which will allow the development of software suitable for the solution of many diverse problems involving differential and partial differential equations.

Differential equations can also be used to describe either the growth or the decay of a population, and as such have many applications in a wide variety of geological problems. Foremost is their application in the study of the kinetics of geochemical processes. A good example which is of interest in this context, is the study of geochemical cycles as formulated by Lasaga (1980, 1981).

Lasaga (1981, p.73) shows that for a two-reservoir geochemical cycle as illustrated in Fig. 8.2, the evolution depends on two rate equations:

$$\frac{dA_1}{dt} = -k_{1,2} A_1 + k_{2,1} A_2$$

$$\frac{dA_2}{dt} = k_{1,2} A_1 - k_{2,1} A_2$$

where $k_{1,2}$ and $k_{2,1}$ are the first order rate constants
and A_1 and A_2 are the amounts of the element in reservoirs 1 and 2 respectively.

The matrix form of these equations is:

$$\frac{d}{dt} \begin{bmatrix} A_1 \\ A_2 \end{bmatrix} = \begin{bmatrix} -k_{1,2} & k_{2,1} \\ k_{1,2} & -k_{2,1} \end{bmatrix} \cdot \begin{bmatrix} A_1 \\ A_2 \end{bmatrix}$$

The solution to this equation is:

$$\mathbf{A}(t) = \mathbf{a}_1 \, e^{\lambda_1 \mathbf{v}_1} + a_2 \, e^{\lambda_2 \mathbf{v}_2}$$

where λ_1 and λ_2 are the eigenvalues of the matrix $\mathbf{K}$
and $\mathbf{v}_1$ and $\mathbf{v}_2$ are the eigenvectors of the matrix $\mathbf{K}$;
$\mathbf{A}(t)$ is a vector of amounts of elements 1 and 2 in reservoirs 1 and 2 respectively, at time (t); a_1 and a_2 are the amounts of the elements in the reservoirs, at time $t = 0$; e is the exponential and $\mathbf{K}$ is the matrix:

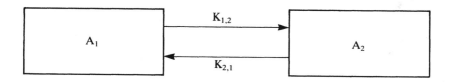

Figure 8.2 A simple two reservoir geochemical cycle. After Lasaga 1981.

$$\begin{bmatrix} -k_{1,2} & k_{2,1} \\ k_{1,2} & -k_{2,1} \end{bmatrix}$$

whose eigenvalues are:

$$\lambda_1 = 0$$
$$\lambda_2 = -(k_{1,2} + k_{2,1})$$

Note: in this application we have departed from the symbols used by Lasaga in favour of the more general mathematical use as given earlier, in Chapter 7.

As Lasaga comments, in order that the cycle returns to a unique steady state one eigenvalue must equal zero (this is a property of all linear cycles), and if the cycle is to be stable then the other eigenvalue must be negative (property of a two reservoir cycle).

This last example illustrates an important application of part of the theory of eigenvalues and eigenvectors, which was not considered in Chapter 7. Students interested in this area of geochemical kinetics, who need to follow up the mathematical basis of this topic, matrix differential equations, should refer to linear algebra texts such as that of Grossman (1984 Ch. 7, p.324ff).

8.3 Well-log analysis and the mixing problem

After this excursion into differential equations and their solution, we now return to something relatively simple, but which from a geological stand-point is no less important. This is the problem of mixing, which we considered early in Chapter 1 in the context of the solution of SLE (section 1.1.1, Example 1.2). The notion of using SLE to solve mixing problems, has been developed by petroleum geologists working in the field of well–log analysis, to enable them to quantify and interpret lithological information present in well–logs. Fundamental to this application is the 'Wylie time-average equation'. This equation assumes that the transit time (as measured by a sonic log), for a sediment with small pores filled with fluid, is related to both lithology and porosity (the equation is not valid if gas is present in the pores). The relationship between transit time and porosity under the conditions defined, is a linear function. Doveton (1986, p.158), writes the equation in matrix form as:

$$\begin{bmatrix} \Delta t_f & \Delta t_{ma} \\ 1 & 1 \end{bmatrix} \cdot \begin{bmatrix} \psi \\ p_{ma} \end{bmatrix} = \begin{bmatrix} \Delta t \\ 1 \end{bmatrix}$$

where ψ = porosity
p_{ma} = proportion of matrix
Δt_f = fluid transit time
Δt_{ma} = transit time through the rock matrix
Δt = observed transit time

Since the response of 'pure' minerals and 'pure' rock to various tools are known, expansion of these equations by including additional information from density and neutron logs, leads to the situation where it is possible to calculate the proportions of rock or mineral types present at any given depth in a borehole, for which these logs are available. Following Doveton (1986, p.159), the matrix equations can be written as:

$$\mathbf{C} \cdot \mathbf{v} = \mathbf{l}$$

which are expanded to give:

$$
\begin{array}{c}
\text{Neutron} \\
\text{Density} \\
\text{Sonic} \\
\text{Unity}
\end{array}
\begin{bmatrix}
\psi & \text{Dolomite} & \text{Sandstone} & \text{Limestone} \\
100 & 13.5 & 3.0 & 10.0 \\
1.0 & 2.683 & 2.485 & 2.54 \\
189.0 & 58.0 & 65.3 & 62.0 \\
1.0 & 1.0 & 1.0 & 1.0
\end{bmatrix}
\cdot
\begin{bmatrix}
\psi \\ D \\ S \\ L
\end{bmatrix}
=
\begin{bmatrix}
x \\ y \\ z \\ 1.0
\end{bmatrix}
$$

where ψ = porosity
D = proportion of dolomite in the observed interval,
S = proportion of sandstone in the observed interval,
L = proportion of limestone in the observed interval,
and x, y and z are readings taken from the Neutron, Density and Sonic logs (respectively), for the observed interval.

The data in the body of the matrix $\mathbf{C}$ are the tool response values taken from tables. To find the vector $\mathbf{v}$ (the unknown proportions of the rock types in the interval and the density), we solve:

$$\mathbf{v} = [\mathbf{C}]^{-1} \cdot \mathbf{l}$$

Once the inverse of the matrix of tool response values has been found, it is a fairly easy and routine matter to calculate the proportions for each lithology, for each interval throughout any length of bore-hole record. Clearly if lithologies other than those given in the matrix above (column labels) are expected, then the matrix will have to be modified accordingly. In practice the situation where the system of equations is under or over determined frequently arises. That is, there are more or fewer lithologies suspected than there are logs available. The reasons for this and possible solutions are discussed in some detail by Doveton, who tabulates equation sets developed, to allow the calculation of 'best estimates' of the proportions under these circumstances (Doveton, 1986, p.177, Table 1).

8.4 Applications in petrological studies

The application of linear algebra to various problems in petrology has been advocated by a number of workers (Perry, 1967; Greenwood, 1975; Thompson, 1982a, 1982b and Spear et al., 1982a, 1982b). The papers by Thompson and Spear et al. outline much of the basic mathematics involved, illustrated by practical examples taken largely from metamorphic petrology. In the field of mineral classification Perry (1967), suggested linear algebra as an alternative approach to the classification of silicate minerals, after investigating their algebraic structure and that of their composition spaces. He suggested that linear algebra could be used to assign minerals to their respective composition spaces with reference to recognized end-members on the basis of their chemical analysis. The assumption is made that the analysis is totally accurate and that the mineral analysed can be regarded as a closed system. With these provisos he suggested that the law of conservation of mass applied to any mineral can be expressed by the following equations (ibid., p.1043):

$$\sum_{j=1}^{n} a_{i,\,j}\, x_j = b_i\ i = 1, 2, \ldots, m$$

where $a_{i,\,j}$ is the number of moles of oxide in molecular end-member j, x_j is the mole per cent of molecular member j in the mineral phase and b_i is the total number of moles of oxide i.

This equation can be written as:

$$\mathbf{A} \cdot \mathbf{x} = \mathbf{b}$$

where $\mathbf{A}$ is a matrix of coefficients relating end-members to oxides, selected so that the end-members are linearly independent. The maximum number of such end-members will be $(m-1)$, where m is the number of oxides.

$\mathbf{x}$ is a vector of proportions of the molecular end-members (mineral composition vector).

$\mathbf{b}$ is a vector of moles of oxide in the mineral analysed.

Although a chemical analysis can always be uniquely expressed in terms of a given set of end-members using this last equation, realistically the calculated proportion of end-members should not be negative; for a discussion on negative components see Thompson (1982a, pp.1–5). This will ensure that the chemical analysis (expressed as vector $\mathbf{b}$) lies within the composition space defined by the chosen end-members. Although it is relatively easy to choose suitable end-members of mineral groups, there is no satisfactory method of determining which are the definitive members to use. For example, in the case of the mica and amphibole groups it is not possible to choose sets of linearly independent end-members (for either group), which will define a composition space that satisfies all possible cases (Perry, 1967, p1077). On the basis of his studies, he showed that silicate minerals fell into one of three groups (p.1052):

(a) groups having a coefficient matrix which contains an $n \times n$ matrix in diagonal form (i.e. there are as many end-members as there are oxides) – the mineral groups are the olivines, feldspars and garnets.
(b) groups having a coefficient matrix which does not contain an $n \times n$ matrix in diagonal form and the number of end-members is less than

the number of oxides (i.e. the system is under determined) – the
mineral groups include the biotites and other micas; and

(c) groups where the number of end-members is greater than the number
of oxides (i.e. the system is over determined) – the amphiboles fall into
this group.

As an example of the problems which can arise, let us consider a member
of group (b), viz. the chlorites. This group of minerals is part of the
phyllosilicate composition space in terms of Perry's classification (ibid.,
p.1044), corresponding to the general formula :

$$(Mg, Al, Fe)_6 (Si, Al)_4 O_{10} (OH)_8$$

(Deer, Howie and Zussman, 1962, p.131). Possible end-members of the
group are:

A $Mg_4 Al_2 Si_2 Al_2O_{10} (OH)_8$ – amesite
B $Mg_6 Si_4O_{10} (OH)_8$ – antigorite
C $Fe^{++}_6 Si_4O_{10} (OH)_8$ – ferroantigorite
D $Fe^{++}_4 Al_2 (Si_2Al_2) O_{10} (OH)_8$ – daphnite

These belong to the orthochlorite group (Hey, 1954). Another group, the
leptochlorites, which are richer in trivalent ions and in particular Fe^{+++},
are also recognized. Two possible end-members being:

E $Fe^{++}_4 Fe^{+++}_2 (Si_2Fe^{+++}_2) O_{10} (OH)_8$
F $Mg_4 Fe^{+++}_2 (Si_2Fe^{+++}_2) O_{10} (OH)_8$

The leptochlorites can be distinguished from the orthochlorites in a
chemical analysis when the level of Fe_2O_3 (if present) is 4% or greater.

Although the above classification might not be universally accepted, it
is sufficiently acceptable to be used to construct a matrix relating
end-members to oxides. This matrix, corresponding to the matrix **A** in
Perry's equation is:

	A	B	C	D	E	F
SiO_2	2	4	4	2	2	2
Al_2O_3	2	0	0	2	0	0
Fe_2O_3	0	0	0	0	2	2
FeO	0	0	6	4	4	0
MgO	4	6	0	0	0	4
H_2O	4	4	4	4	4	4

Following the rules given above, the selection of $(m-1)$ linearly independent end-members should be made from the columns of this matrix. This
will lead to a matrix with more rows (6 oxides) than columns (maximum
5 end-members), in which case it will not be possible to solve the
equation:

$$[A]^{-1} \cdot x = b$$

However, for the chlorites the problem does not end here. As many analyses are now routinely carried out using a microprobe, data relating to Fe_2O_3 and H_2O are not directly available, thus the foregoing matrix can be reduced (for microprobe analysis in particular), by excluding these two oxides as well as the two end-members representing the leptochlorites. This leads to a 4×4 matrix relating the four orthochlorites to the oxides SiO_2, Al_2O_3, FeO and MgO. This is:

$$
\begin{array}{cccc}
 & A & B & C & D \\
SiO_2 & \begin{bmatrix} 2 & 4 & 4 & 2 \\ Al_2O_3 & 2 & 0 & 0 & 2 \\ FeO & 0 & 0 & 6 & 4 \\ MgO & 4 & 6 & 0 & 0 \end{bmatrix}
\end{array}
$$

As these four end-members are not themselves linearly independent, the next stage of the process is to take combinations of three end-members and their corresponding oxides. In this situation, since there are more oxides than end-members, it is possible to check legal assemblages by using different combinations of oxides with the same group of end-members.

The combinations of the four end-member minerals which are possible are:

$$
\begin{bmatrix}
A & B & C & + \\
A & B & + & D \\
A & + & C & D \\
+ & B & C & D
\end{bmatrix}
$$

Since all the minerals contain SiO_2 it would be convenient to base any assignments on the oxides Al_2O_3, FeO and MgO. If this were carried out, all but one of the four combinations can be utilized, the exception being: $B - C - D$ ($Det[A] = 0$). An additional bonus gained by using this strategy, is that legal assignments could be checked using combinations of these three oxides with SiO_2.

Having argued the advantages of using matrix algebra to tackle problems of silicate geochemistry, we have clearly demonstrated a major problem. Does the difficulty lie in the recognition and assignment of mineral end-members, or does it lie in the algebra? Obviously much further work is required before a satisfactory answer to this dilemma can be given, sadly this is beyond the scope of this book.

Returning to other applications of linear algebra in petrology, extensive studies have been made in the field of metamorphism. In this context the *Mineralogical Society of America Review*, Volume 10 (1982), contains much of interest. The first three chapters of this work (referred to earlier in this section), dealing with composition space, are particularly relevant.

This volume as with others in the series, contains extensive lists of references detailing earlier applications.

In geochemical thermodynamics the 'linearity' of chemical reactions is exploited in Gibbs free energy determinations (section 1.2.1, Example 1.5 gives a simple application). Since this is utilized by geologists to spread errors in thermodynamic data by considering a number of reactions simultaneously, it is worthwhile considering in more detail. The method, known as **simultaneous retrieval**, starts by setting up a series of reactions involving minerals with 'known' and 'unknown' Gibbs free energy. By using tabulated data from different sources, for minerals with 'known' values; the unknowns can be calculated and discrepancies due to differences in accuracy in the original sources can be spread (Nordstrom and Munoz, 1986, p.417ff).

Before considering a specific example, and for those without the necessary thermodynamic background the following discussion might be helpful:

Gibbs free energy (usually denoted by the symbol G), is an extensive property of a chemical system and can be considered as the available energy, tending to a minimum value at equilibrium. Thus for any natural process to occur spontaneously it must be accompanied by a fall in G, hence the loss of free energy $(-\Delta G)$ measures the thermodynamic tendency for the reaction to proceed. If the process is spontaneous and natural ΔG is large and negative.

There are two measures of free energy:

$$\Delta G^\circ_f - \text{the free energy of formation}$$
$$\Delta G^\circ_r - \text{the free energy of reaction}$$

The units of measurement are: $K j \, mol^{-1}$ or $cal \, mol^{-1}$
the latter can be found in older texts.

Consider the reaction:

$$3Fe_2O_3 + CO = 2Fe_3O_4 + CO_2$$

Now if we are given the Gibbs free energy of formation of the four phases in the reaction, the Gibbs free energy for the reaction can be calculated:

	$\Delta G^\circ_f \, (k j \, mol^{-1})$
Fe_2O_3	−743.6
Fe_3O_4	−1017.1
CO_2	−394.4
CO	−137.2

Then, remembering that reactants are negative and products are positive:

$$\Delta G^0_r = -3(-743.6) - (-137.2) + 2(-1017.1) + (-394.4)$$
$$= 2368.0 - 2428.6$$
$$= -60.6 \text{ K j mol}^{-1}$$

Further, if we now consider the reactions involved in the production of iron from its ore, which are:

$$\Delta G^0_r \text{ (K j mol}^{-1})$$

$$
\begin{array}{ll}
3Fe_2O_3 + CO = 2Fe_3O_4 + CO_2 & -60.6 \\
Fe_3O_4 + CO = 3FeO + CO_2 & +5.4 \\
FeO + CO = Fe + CO_2 & -5.7
\end{array}
$$

In this set of reactions there are 3 solid phases, 2 gaseous phases plus the element itself. Now, since the ΔG^0_f of an element is zero, we can find any three ΔG^0_f for these reactions, provided we are given the other two. This follows since, if we treat the equations algebraically, they are under determined with 3 equations and 5 unknowns:

$$
\begin{array}{rcl}
-3x_1 - x_2 + 2x_3 + x_4 & = & -60.6 \\
-x_3 - x_2 + 3x_5 + x_4 & = & 5.4 \\
-x_5 - x_2 + x_4 & = & -5.7
\end{array}
$$

(Note signs, as given earlier)
Suppose we are given:

$$\Delta G^0_f \ CO_2 = -394.4 \ (x_4)$$
$$\Delta G^0_f \ CO = -137.2 \ (x_2)$$

Then the equations become:

$$
\begin{array}{rcl}
-3x_1 - (-137.2) + 2x_3 + (-394.4) & = & -60.6 \\
-x_3 - (-137.2) + 3x_5 + (-394.4) & = & 5.4 \\
-x_5 - (-137.2) + (-394.4) & = & -5.7
\end{array}
$$

These three equations can now be solved to give ΔG^0_f for the three oxides of iron:

$$
\begin{array}{ccc}
Fe_2O_3 & Fe_3O_4 & FeO \ . \\
\end{array}
\ \ \Delta G^0_f \ \ = \ \ \Delta G^0_r
$$

$$
\begin{bmatrix}
-3 & 2 & 0 \\
0 & -1 & 3 \\
0 & 0 & -1
\end{bmatrix}
.
\begin{bmatrix}
Fe_2O_3 \\
Fe_3O_4 \\
FeO
\end{bmatrix}
=
\begin{bmatrix}
196.6 \\
262.6 \\
251.5
\end{bmatrix}
$$

Since the matrix is in upper diagonal form, back substitution leads to:

$$\Delta G^0_f \ Fe_2O_3 = -743.6 \text{ (K j mol}^{-1})$$
$$\Delta G^0_f \ Fe_3O_4 = -1017.1 \text{ (K j mol}^{-1})$$
$$\Delta G^0_f \ FeO = -251.5 \text{ (K j mol}^{-1})$$

It should be appreciated that the choice of reactions is important, since the matrix formed from the reactants and products must not show any linear dependencies i.e. the reactions themselves must be linearly independent.

8.5 Tectonic studies

The importance of linear algebra in certain limited fields within the discipline of structural geology, has been demonstrated in Chapters 5 and 6. Unfortunately it is not practical to carry out a complete survey of other structural applications, or to review the very extensive literature on the subject at this juncture. Students are reminded that the text books of Ramsay and Huber (1983, 1987) noted earlier, cover much that is relevant in this field, as well as containing excellent annotated references.

Before moving on however, one further area of study is worth mentioning, since it considers the problems and limitations of working in two dimensions. Wheeler (1986) presents a study in strain-analysis, considering the relatively common geological situation where deformed ellipsoid markers had some initial preferred orientation. In this area several approaches have been made, ranging from those dependent purely on statistics, to those involving algebraic arguments. As a result of his studies, Wheeler outlines a simple solution to the problem involving linear algebra, while at the same time demonstrates that the more frequently practised two-dimensional approach, is less than satisfactory in many situations. The equations developed and used in the paper are fully detailed in the appendices (Wheeler, 1986, p.894ff).

Much of the subject matter considered in the earlier chapters has been concerned with points and lines in the plane and in space. In the examples used to demonstrate the various techniques considered, we have expressed

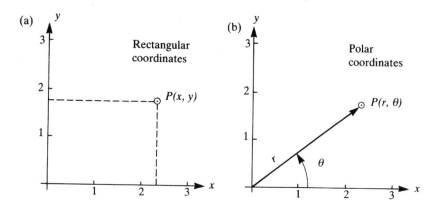

Figure 8.3 The relationship between rectangular coordinates and polar coordinates.

the coordinates in terms of either a rectangular coordinate system (for example trend surface analysis) or a polar coordinate system (primarily in the context of the magnitude and direction of vectors), without considering how the systems are inter-related. To remedy this shortcoming consider Fig. 8.3 which demonstrates the differences between the two systems in 2-dimensions. From this it should be clear that to convert from one system to the other, we can use the equations:

$$x = r \cos \theta \; ; \; y = r \sin \theta$$

or

$$r = \sqrt{(x^2 + y^2)} \; ; \; \tan \theta = \tfrac{x}{y}$$

In three dimensions these systems correspond to the cartesian and spherical coordinate systems, respectively. The convention adopted for the three mutually perpendicular axes is in the upper northern hemisphere, with reference to Fig. 8.4, is:

x – will be in the plane of the equator toward $\phi = 0°$
y – will also be in the plane of the equator, but toward $\phi = 90°$
z – will be aligned with its rotational axis toward the north Pole

The appropriate equations, resolving the unit vector $a = (\lambda \; \phi)$ are:

$$\mathbf{a}_{\text{polar}} = \sin \lambda$$
$$\mathbf{a}_{\text{equatorial}} = \cos \lambda$$

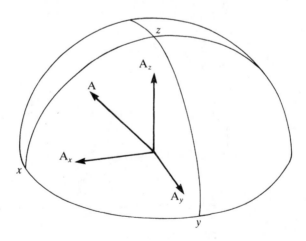

Figure 8.4 The convention adopted for three mutually perpendicular axes, for spherical coordinates.

Thus for cartesian components:

$$A_x = \cos \lambda \cos \phi$$
$$A_y = \cos \lambda \sin \phi$$
$$A_z = \sin \lambda$$

and converting back to spherical:

$$\lambda = \text{arc sin } A_z$$

$$\phi = \text{arc tan } \left(\frac{A_y}{A_x}\right)$$

where A_x, A_y and A_z are the coordinates of A, and arc sin is a function returning 'the angle whose sine is', sometimes written $\sin^{-1}\phi$; similarly arc tan.

Also since **a** is a unit vector, then:

$$A_x^2 + A_y^2 + A_z^2 = 1$$

and only two of the three cartesian components are independent (Cox and Hart, 1986).

Clearly in the study of plate tectonics, relative movements of areas of the earth's surface in time and direction are important. In this context it is necessary to produce maps of the earth's surface which show movement in terms of rotation and velocity components. These can be produced either by geometric or algebraic means and Cox and Hart (1986), in their excellent book on the mechanism of plate tectonics give the matrix equations necessary for the algebraic approach. This emphasizes the fact that much of linear algebra is concerned with transforming data from one coordinate system to another. Thus in many of the applications in mineralogy (particularly the work of Perry referenced earlier), we are simply converting from a coordinate system based on chemical analysis (usually oxides) into another system based on mineral end-members. What is important, whether we are concerned with events on a chemical scale or on a global scale, is that in many instances a considerable economy is achieved both in the representation of a problem and in its solution, when it is formulated in terms of linear algebra.

Finally it is worth mentioning that in this book the 'linear' in linear algebra has been taken quite literally and situations where non-linear terms could be involved has been largely ignored. Two examples spring to mind. In structural geology only homogeneous deformation has been considered and the more realistic heterogeneous deformation has been ignored. In many real situations deformation changes with time, often continuously, hence in the definition of heterogeneous deformation the individual terms in the transformation matrix can be differentials (see Ramsay and Huber, 1983, p.33ff). In trend surface analysis we considered the linear situation in detail, only mentioning in passing that the method could be extended to include higher-order surfaces. Indeed it is imperative in practice, to use such higher-order surfaces. In this situation, the linear equations can be expanded to allow coefficients of the higher order terms to be calculated; see Davis (1973, p.331ff). The mathematical basis (minimizing the sums of squares of the deviations from the trend), using partial differentials to arrive at the SLE set, is very similar to that used in Chapter 1.

Answers to selected student examples

Chapter 1

Q1.1   3 motor cycles; 17 cars

Q1.2   Treak Cliff vs Thorpe Cloud: $W = 16.8$ mm; $L = 12.9$ mm
Treak Cliff vs Bolland: $W = 28.9$ mm; $L = 20.9$ mm
Thorpe Cloud vs Bolland: $W = 10.3$ mm; $L = 7.9$ mm

Q1.3   methane $= 794$ vppm; ethane $= 5.77$ vppm

Q1.4 $a = 0.19$; $c = 8.31$ or $Pb = 0.19\ Zn + 8.31$

Q1.5   2 enstatite + 3 periclase = 2 forsterite + FeO

Q1.6 $x = 1$; $y = 1$; $z = 1$

Q1.7 $x_1 = 1$; $x_2 = -1$; $x_3 = 2$; $x_4 = -2$

Q1.8   (a) $x_1 = -0.5$; $x_2 = -1.5$
(b) $x_1 = 2.5$; $x_2 = -3.0$

Q1.10 grossular + quartz = 2 wollastonite + anorthite

Q1.11 Equations are not linearly independent

Q1.13 (a) $[-2, -1, 1]$; (b) $[1, 1, -1]$

Q1.14 5 garnet + chlorite + 8 muscovite =
                        4 biotite + 16 kyanite + 8 water

Chapter 2

Q2.1   (a) (1.16 2.9 3.48 5.22); (b) $(0\ \ -1.5B\ \ 2.3B\ \ 4.0B)$
(c) (7.5 5 12.5 22.5); (c) $(-27\ \ -6\ \ 9\ \ 12)$

Q2.2   (a) 20; (b) 16; (c) 15; (d) -2

Q2.3 $(\mathbf{a.b}) = -20$; $(\mathbf{b.a}) = -20$
Therefore $(\mathbf{a.b}) = (\mathbf{b.a})$

Q2.5   (a) $\begin{bmatrix} 25 \\ 21 \\ 20 \end{bmatrix}$ (b) $\begin{bmatrix} 4 \\ 25 \\ 11 \end{bmatrix}$

Q2.6   **W**(i) on the basis of 24 oxygens:

$$\begin{array}{l} \text{Si} \\ \text{Al} \\ \text{Fe}^{+++} \\ \text{Fe}^{++} \\ \text{Mg} \\ \text{Ca} \\ \text{Na} \\ \text{H}_2 \\ \text{O} \end{array} \begin{bmatrix} 7.1961 \\ 1.2143 \\ 0.2630 \\ 0.6180 \\ 3.7576 \\ 1.8398 \\ 0.1641 \\ 1.0744 \\ 24.0000 \end{bmatrix}$$

Q2.7 $[\mathbf{A}] \cdot [\mathbf{B}] = \begin{bmatrix} 16 & 9 & 7 \\ 28 & 14 & 12 \\ 16 & 5 & 6 \end{bmatrix}$ $[\mathbf{B}] \cdot [\mathbf{A}] = \begin{bmatrix} 11 & 5 & 13 \\ 22 & 10 & 20 \\ 15 & 7 & 15 \end{bmatrix}$

Therefore: $[\mathbf{A}] \cdot [\mathbf{B}] \neq [\mathbf{B}] \cdot [\mathbf{A}]$

Q2.8   (a) $[\mathbf{A}] \cdot [\mathbf{B}] = \begin{bmatrix} 22 & 20 \\ 20 & 25 \end{bmatrix}$ $[\mathbf{B}] \cdot [\mathbf{A}] = \begin{bmatrix} 29 & 22 & 11 \\ 9 & 14 & 4 \\ 10 & 12 & 4 \end{bmatrix}$

(b) $[\mathbf{A}] \cdot [\mathbf{B}] = \begin{bmatrix} 37 & 11 & 10 \\ 27 & 12 & 10 \end{bmatrix}$ $[\mathbf{B}] \cdot [\mathbf{A}]$ not possible since $j \neq 1$

Q2.10 $\begin{bmatrix} 92 & 41 & 68 \\ 119 & 52 & 81 \\ 83 & 36 & 56 \end{bmatrix}$

Chapter 3

Q3.1   (a) 2; (b) −2; (c) −42; (d) 0; (e) 13

Q3.2   (a) Det = 1 – exchanging rows (2) and (3) Det = −1, then: exchanging cols (2) and (3) Det = 1

(b) Det = −3 – exchanging cols (1) and (3) Det = 3, then: exchanging rows (1) and (3) Det = −3

Q3.5   Det = 15 therefore the equations are linearly independent

Q3.6   (a) Det = 13 Inverse = $\begin{bmatrix} 0.385 & 0.077 \\ 0.154 & 0.291 \end{bmatrix}$

(b) Det = 9 Inverse = $\begin{bmatrix} 0.889 & -0.778 \\ -0.111 & 0.222 \end{bmatrix}$

(c) Det = 7 Inverse = $\begin{bmatrix} 0.286 & 0.143 \\ 0.429 & 0.714 \end{bmatrix}$

(d) Det = −9 Inverse = $\begin{bmatrix} \frac{20}{27} & \frac{-5}{9} & \frac{-1}{9} \\ \frac{-10}{9} & \frac{11}{9} & \frac{4}{9} \\ \frac{-2}{9} & \frac{4}{9} & \frac{-1}{9} \end{bmatrix}$

Q3.7   Det = 0 , No inverse

Q3.8   Det = -1 , $x_1 = 2$; $x_2 = 3$; $x_3 = 4$

Chapter 4

Q4.2 $x_1 = 2.0514$; $x_2 = 1.2509$; $x_3 = 1.4319$ (without pivoting)
        $x_1 = 2.0503$; $x_2 = 1.2626$; $x_3 = 1.4289$ (with pivoting)

Q4.4 $x_1 = 2.1972$; $x_2 = -0.1487$; $x_3 = 1.7884$

Chapter 5

Q5.1   (a) Length = 2.2361 Direction = 63° 27'
        (b) Length = 9.434 Direction = 122°
        (c) Length = 7.2111 Direction = 56° 18'
        (d) Length = 3.6056 Direction = 326° 18'

Q5.3   (a) Length = 4.5826 Direction = (64° 6', 77° 24', 34° 12')
        (b) Length = 2.2913 Direction = (64° 6', 34° 12', 77° 24')

Q5.4   Vector pairs (a), (d), (e), (g) & (h) are mutually orthogonal
        Vector pairs (b), (c) & (f) are not mutually orthogonal

Q5.5   The equations of the 3 lines are:
        $y = 1.43x_1 - 2.0$
        $y = 0.93x_2 + 1.5$
        $y = 0.43x_3 + 5.0$
        and they meet at the point (7, 8)

Q5.6   The limestone has a plane surface

Q5.9 $[A]^4 = \begin{bmatrix} 41 & 40 \\ 40 & 41 \end{bmatrix}$ and the new position = $\begin{bmatrix} 81 \\ 81 \end{bmatrix}$

Q5.11 & Q 5.12 See Chapter 6 (beginning).

Chapter 6

Q6.1   (a) $\lambda_{max} = 6.6055$ $v_{\lambda_{max}} = (1\ 1.151)$
            $\lambda_{min} = 0.6055$ $v_{\lambda_{min}} = (1\ -0.651)$
            Not orthogonal
        (b) $\lambda_{max} = 1.43$ $v_{\lambda_{max}} = (1\ 1.075)$
            $\lambda_{min} = 0.07$ $v_{\lambda_{min}} = (1\ -2.326)$
            Not orthogonal

Q6.2 $-\lambda^3 + 2\lambda^2 - 10.21\lambda + 5.42$

Q6.3   For pure shear $\lambda_{max} = 2$, $\lambda_{min} = 0.5$

For simple shear $\lambda = 1.0$ (roots coincident)
In both cases the eigenvectors are parallel to the axes

Q6.5 $\lambda_{max} = 4.5616$ $v_{\lambda_{max}} = (1\ 0.6404)$

Q6.6 $\lambda_{max} = 4.0$ $v_{\lambda_{max}} = (1\ 1\ 1)$

Q6.7 $\lambda_{max} = 7.52$ $v_{\lambda_{max}} = (1\ 0.49\ 0.85)$

Q6.10 For the matrix: $\begin{bmatrix} 1.2 & 0.2 \\ 0.4 & 0.8 \end{bmatrix}$

$\lambda_{max} = 1.35$ $v_{\lambda_{max}} = (1\ 0.73)$
$\lambda_{min} = 0.65$ $v_{\lambda_{min}} = (1\ -2.73)$
Length of the principal axis = 1.37
Length of the minor axis = 0.64
Direction before strain 31°
Direction after strain 25°10′
Rotation 5°43′
For the matrix: $\begin{bmatrix} 1 & 0.6 \\ 0.6 & 1.5 \end{bmatrix}$

$\lambda_{max} = 1.9$ $v_{\lambda_{max}} = (1\ 1.5)$
$\lambda_{min} = 0.6$ $v_{\lambda_{min}} = (1\ -0.667)$
Length of the principal axis = 1.9
Length of the minor axis = 0.6
Zero rotation

Q6.11 (a) Percentage difference 0.398%
(b) Difference in slope 5° 10′
(c) Rotational component 0° 57′ clockwise

Chapter 7

Q7.1 Sums of squares, sums of cross products matrix:
$$\begin{bmatrix} 1186.410 & 385.260 & 8.223 \\ 385.260 & 125.220 & 2.669 \\ 8.223 & 2.669 & 0.057 \end{bmatrix}$$

Dispersion matrix:
$$\begin{bmatrix} 0.809 & 0.037 & 0.011 \\ 0.037 & 0.054 & 0.001 \\ 0.011 & 0.001 & 0.0001 \end{bmatrix}$$

Q7.2 Sums of squares, sums of cross products matrix:
$$\begin{bmatrix} 220.461 & 259.487 & 21.475 \\ 259.487 & 829.290 & 78.729 \\ 21.475 & 78.729 & 8.045 \end{bmatrix}$$

Dispersion matrix:

$$\begin{bmatrix} 68.594 & -65.912 & -9.701 \\ -65.912 & 132.065 & 11.929 \\ -9.701 & 11.929 & 1.645 \end{bmatrix}$$

Q7.3   Thickness = 0.923 Length + 0.003 Width $-$ 0.330

Q7.4   The linear trend surface equation is:
$z_{trend} = 0.12997 + 0.041x - 0.011y$
where x = easting, y = northing, z = thickness

Q7.5 R_{Grp1} = 80.983, R_{Grp2} = 46.593, R_o = 63.788
R_{s1} = 33.876 = Group 2
R_{s2} = 11.255 = Group 2
R_{s3} = 97.265 = Group 1
R_{s4} = 43.925 = Group 2
R_{s5} = 80.580 = Group 1

Q7.6   Dispersion matrix:
$$\begin{bmatrix} 545.765 & -822.882 \\ -822.882 & 6992.941 \end{bmatrix} \begin{matrix} (Zn) \\ (Cu) \end{matrix}$$

Principal Components Analysis:

	Evect 1	Evect 2
Variable 1 (Cu)	−0.125	0.992
Variable 2 (Zn)	0.992	0.125
Eigenvalue	7096.312	442.394
Variation accounted	92.76%	7.24%

References

Battey, M.H., (1981) *Mineralogy for students* (2nd edn) Longmans, London.

Boas, M.L., (1983) *Mathematical methods in the physical sciences* (2nd edn) John Wiley, New York.

Cox, A. and Hart, R.B., (1986) *Plate tectonics, how it works*, Blackwell Scientific Publishers, Oxford.

Davis, J.C., (1973) *Statistics and data analysis in geology*, John Wiley, New York.

Deer, W.A., Howie, R.A. and Zussman, J., (1962) *Rock-forming minerals, Vol. 3: Sheet Silicates.* Longmans, London.

Doveton, J.H., (1986) *Log analysis of subsurface geology*, Wiley-Interscience, New York.

Ferguson, J., (1991) The organic geochemistry of hydrocarbon gases in flourites from northern England, *J. Petroleum Geol.*, **14**, 221–8.

Ferry, J.M. (ed), (1982) Characterization of metamorphism through mineral equilibria, *Reviews in Mineralogy, Min. Soc. Amer.*, **10**.

Graham, D., Graham, C. and Whitcombe, A., (1984) *A-level mathematics course companion,* Letts Study Aids, Charles Lett, London.

Gray, J.R., (1967) *Probability*, Oliver and Boyd, Edinburgh.

Greenwood, H.J., (1975) Thermodynamically valid projections of extensive phase relationships. *Amer.Mineral.*, **60**, 1–8.

Grossman, S.I., (1984) *Elementary Linear Algebra*, (2nd edn) Wadsworth Publishing Co., Belmont, California.

Harbaugh, J.W. and Merriam, D.F., (1968) *Computer applications in stratigraphic analysis*, John Wiley, New York.

Harbaugh, J.W. and Bonham-Carter, G., (1970) *Computer simulation in geology*, Wiley-Interscience, New York.

Hey, M.H., (1954) A new review of the chlorites, *Min. Mag.*, **30**, 277–92.

Lasaga, A.C., (1980) The kinetic treatment of geochemical cycles, *Geochim. Cosmochim. Acta*, **40**, 257–66.

Lasaga, A.C., (1981) Dynamic treatment of geochemical cycles: Global kinetics, in *Kinetics of geochemical processes*, (eds A.C. Lasaga and R.J. Kirkpatrick) *Reviews in Mineralogy, Min. Soc Amer.*, **8**, 69–110.

Lipschultz, S., (1968) *Schaum's outline of theory and problems in linear algebra*, SI (metric) edn, McGraw-Hill Book Co., New York.

Mathews, J.H., (1992) *Numerical methods for mathematics, science and engineering*, (2nd edn) Prentice-Hall International, Englewood Cliffs, New Jersey.

Moroney, M.J., (1953) *Facts from figures* (2nd edn) Penguin Books, London.

Nordstrom, D.K. and Munoz, J.L., (1986) *Geochemical thermodynamics*, Blackwell Scientific Publishers, Oxford.

Parkinson, D., (1969) Relative growth, variation and evolutionary trends in a Carboniferous rhynchonellid brachiopod, *J. Paleont.*, **43**, 95–110.

Perry, K. Jr., (1967) An application of linear algebra to petrological problems: Part 1 Mineral classification, *Geochim. Cosmochim. Acta*, **31**, 1043–78.

Ragan, D.M., (1973) *Structural Geology an introduction to geometrical techniques* (2nd edn) John Wiley, New York.

Ramsay, J.G. and Huber, M.I., (1983) *The techniques of modern structural geology. Vol. 1 Strain analysis*, Academic Press, London.

Ramsay, J.G. and Huber, M.I., (1987) *The techniques of modern structural geology. Vol. 2 Folds and fractures*, Academic Press, London.

Sieveking G. de G., Craddock P.T., Hughes M.J., Bush P. and Ferguson, J., (1970) Characterization of prehistoric flint mine products, *Nature*, **228**, 251–4.

Sieveking G. de G., Bush P., Ferguson, J., Craddock P.T., Hughes M.J. and Cowell, M.R., (1972) Prehistoric flint mines and their identification as sources of raw material, *Archaeometry*, **14**, 151–76.

Spear, F.S., Rumble, D.III and Ferry, J.M., (1982a) Linear algebraic manipulation of N-dimensional composition space, *Reviews in Mineralogy*, Vol. 10, (J.M. Ferry, ed) *Characterization of metamorphism through mineral equilibria*, Min. Soc. Amer., 53–104.

Spear, F.S., Rumble, D.III and Ferry, J.M., (1982b) Analytical formulation of phase equilibria: The Gibbs method. *Reviews in Mineralogy*, Vol. 10, (J.M. Ferry, ed) *Characterization of metamorphism through mineral equilibria*, Min. Soc. Amer., 105–52.

Stephenson, G., (1973) *Mathematical methods for science students* (2nd edn) Longman, London.

Thompson, J.B., jr., (1982b) Reaction space: An algebraic and geometric geometric approach. *Reviews in Mineralogy*, Vol. 10, (J.M. Ferry, ed) *Characterization of metamorphism through mineral equilibria*. Min. Soc. Amer., 1–31.

Thompson, J.B.,jr., (1982b) Reaction space: An algebraic and geometric approach. *Reviews in Mineralogy*, Vol. 10, (J.M. Ferry, ed) *Characterization of metamorphism through mineral equilibria*. Min. Soc. Amer., 33–52.

Wheeler, J., (1986) Strain analysis in rocks with pre-tectonic fabrics. *J. Struct. Geol.*, **8**, 887–96.

Windle, A., (1977) *A first course in crystallography*, Bell, London.

Index